2"x6"x2'
Bevel top for safety, or cap with wooden or aluminum top
Holes every 3 inches
Nails
bridle Hook
handle
$6^1/_2$"
$3^1/_2$"
$2^1/_2$"
$5^1/_4$"
20"
BACK
$21^1/_2$"h x 24" w
each step: $10^1/_2$" x 24"
each step face: $6^1/_4$" x24"
top: 11" 24"
front: 7" x 24"
TOP
FRONT
11"
$10^1/_2$"
$6^1/_4$"
$21^1/_2$"
Handle on each side
30"
($31^1/_8$" total length)
24"
7"
2"x4"x12'
2'x8' 5/8" plywood
2'x3'4"
1"x4"x11'4"
31/2"x4" filler
Measure out and
10"
drill
21"
28"
To conserve plywood, first cut rectangle to 28" x 30"

Jumps, etc.

Lisa Campbell

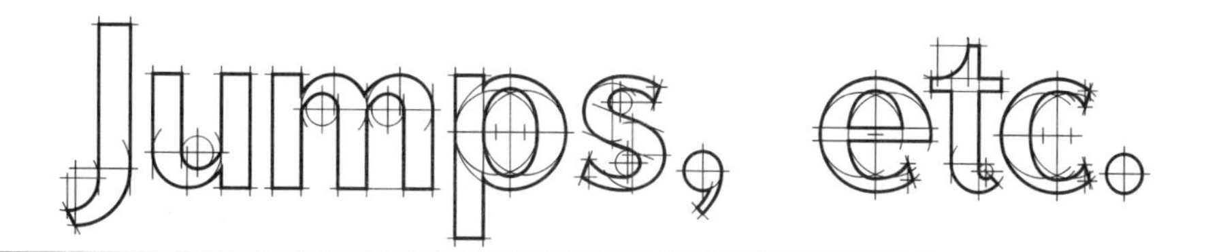

Jumps, Dressage Arenas and Stable Equipment **You Can Build**

Illustrations by Diann Landau
Photos by Lisa Campbell

Boonsboro, Maryland

Jumps, etc.

Jumps, Dressage Arenas and Stable Equipment
You Can Build

Published in the United States of America by

Half Halt Press, Inc.
P.O. Box 67
Boonsboro, MD 21713

Book and cover design by Design Point

Printed in the United States of America

Library of Congress Cataloging-in-Publication Data
Campbell, Lisa, 1954-
Jumps, etc. : jumps, dressage arenas, and stable equipment you can build / Lisa Campbell ; illustrations by Diann Landau ; photos by Lisa Campbell.
p. cm.
Includes bibliographical references (p.).
ISBN 0-939481-56-1 (hardcover)
1. Courses (Horse sports)--Design and construction. I. Title.

SF294.35 .C36 2000
688.7'8--dc21 00-039560

Dedication

This book is dedicated to my mother, Ruth Williams, who encouraged me to write since I was a child, and to my late father, Harold Williams, who taught me many practical things in life including how to use tools, build things, and change the oil in the car, among an endless list of other life skills.

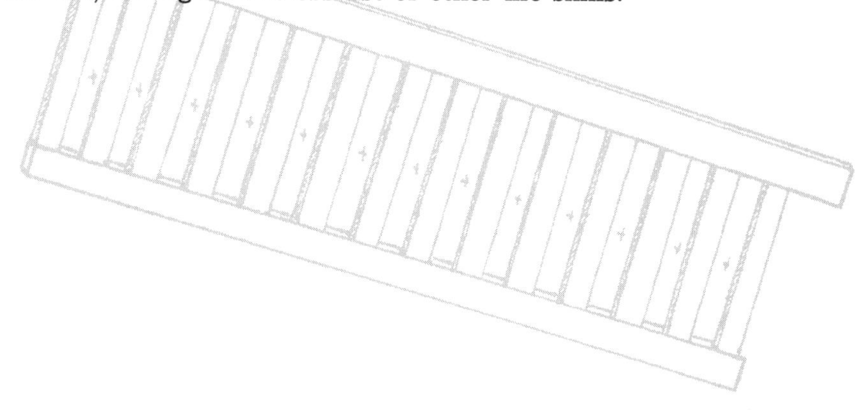

Table of Contents

Table of Contents
continued

Acknowledgments

This book could not have been written without the help of my husband, Dave Campbell, who helped design most and built all of the jumps in this book. Many plans are based on drawings that originated at Virginia Polytechnic Institute and State College in Blacksburg, Virginia, years ago, but we have modified them to fit modern jump designs and to use industry standard lumber available at lumber yards and home improvement stores.

Carol Noggle, former D.C. of Dominion Valley Pony Club (Va.), patiently taught me how to set up a dressage arena and shared with me what materials to use for the border. The letters in Chapter 7 and the tack box/saddle stand in Chapter 4 are Dominion Valley's design.

Diann Landau, who penned all the line drawings, is an accomplished horsewoman, instructor and trainer. She brought her vast horse knowledge willingly to this project.

Some of the material in this book was originally published in *The Chronicle of the Horse*. My thanks to Rob Banner, my publisher at *The Chronicle*, and John Strassburger, my editor, both of whom granted me permission to reprint it.

Introduction

Riders who enjoy the thrill of riding a well-trained horse over jumps may some day want practice jumps in the home arena or pasture. Trainers of green horses need basic training fences and later some additional but more interesting jumps to give their mounts a proper foundation of both physical and mental preparation for the show ring, combined training course or the hunt field. Boarding stable managers and riding instructors need jumps for their clients to practice riding over. Perhaps he or she may find themselves organizing a new horse show and need a good solid course of jumps.

Buying a small course of a few practice jumps represents a sizable investment, just like buying a good saddle. But while saddles are easy to find in tack stores, jumps are often unavailable for purchase locally. Buying a course of jumps from an out-of-state jump manufacturer will incur exorbitant shipping costs because jumps are large and heavy items.

But anyone with the ambition and a few good tools can make sturdy attractive jumps on his or her own. This book is intended for just that person, a neophyte carpenter, the person who knows just a little about carpentry to give it a try. With a safe, sound power saw, an electric drill, a hammer and a can-do spirit, the plans in this book will pave the way.

Should the reader decide the project is too overwhelming or time-consuming, the plans contained in Chapter 4 can be handed to a professional carpenter or an enthusiastic spouse or friend who is an amateur carpenter, even those who know nothing about horses and even less about horse jumps! Landau's professional line drawings will guide the carpenter who normally works from blueprints.

Chapter One

Determining Your Jump Needs

Planning what jumps to build is similar to deciding how to furnish your house. There are many different types intended for different uses, and the amount of "floor space" available is a limiting factor. The questions to consider of how many and what kind of jumps involve the arena space and the planned use of the jumps. These questions include:

- ✔ Will the jumps be used in an arena or in a field?
- ✔ What size arena will the jumps be used in?
- ✔ Will the jumps be used for schooling only or will they be used for shows?
- ✔ Will the primary users be inexperienced or advanced riders or horses?

The area available for jumping

Before deciding how many jumps and what size to make them, consider what they are to be used for and make a careful assessment of the area available to set up the jumps. With a large area in which to school, a progressive series of jumps can be set up. The rider can start small for the warm-up, and then progress to larger or more complex obstacles without dismounting to adjust heights.

However, much can be done in a small area with four to six basic schooling standards and 10 to 14 poles.

If you are planning to build new jumps for a show jumping or hunter course, first consult with the course designer to produce jumps that are suitable for the arena space and meet the rules of the competition.

If using a large open field, 12-foot long poles and other corresponding equipment will make a bigger target, an important consideration if schooling green horses or riders. Out in the open, horses will naturally take bigger strides and move in a more forward manner than in an arena. A large open space can naturally accommodate more jumps than riding arenas limited by fencing or walls. Several jumps can be set up for progressive schooling, from a basic warm-up cross-pole jump to a full course.

If using a large riding arena, indoor or outdoor, consider how many comfortable canter strides a horse can take on the long side when cantering at a suitable pace for jumping. That will give you an idea of how many jumps can go on the long side. If green horses or riders are schooling, will they be able to jump a line of two or three jumps, or a grid, and then make the turn on the short side without scrambling for balance?

When considering 10-foot versus 12-foot poles, remember that two 12-foot pole jumps will take up an additional four feet of arena width. Room should be left for the horse to comfortably school between the jumps, and between the jumps and arena walls or railing. The arena width should also be considered when choosing between schooling standards or wing standards. The base of wing standards are 30 inches each, which will add at least 60 inches. The base of schooling standards are 2 feet wide.

If using a small riding arena, a basic set of jumps which includes four to six pair of schooling standards and 10-foot poles and obstacles may be all that will fit in an arena the size of a standard dressage arena (20 mm x 60 mm). With these basic jumps, the trainer can make four jump combinations, two of which can be oxers, or a gymnastic grid can be set up. Many variations are possible with just a few jumps.

Use of the Jumps

The next thought to mull over concerns the actual use of the jumps. Will they be for starting young horses over fences aimed for the 3-foot hunter ring, or for schooling horses destined for the Grand Prix ring? Will the jumps be used for lessons limited to students just beginning to jump, or will your lessons include advanced jumping?

If the jumps are for schooling only, will the primary users be adults or children? Who will be moving the equipment around to reset the course?

Are the jumps to be used in horse shows? Will the shows be AHSA recognized or will they be local schooling shows? Will the shows be strictly for hunters or will jumper classes be offered? Are the jumps to be used for a USCTA show jumping phase?

All of these uses have particular requirements that should be considered.

Jumps For Horse Shows

If the jumps will be used for horse shows, the first task is to obtain a rule book for the organization governing the show. Refer to the rules concerning the parameters of the course and the design for the classes that will be offered before shopping for materials.

Breed shows may have rules close to the AHSA rules, but to be safe, check with the rules covering the specific breed. If building jumps for an Arabian jumper class, consult the IAHA rule book, or if building jumps for a Quarter Horse jumper class, consult the AQHA rule book, etc.

After determining who will be using the jumps, you'll need to select a suitable size for obstacles such as panels, gates, pickets or coops for the users—ponies or jumpers, recognized or unrecognized shows, or hunters, jumpers or combined training. If the jumps will be used in recognized AHSA hunter classes, the jumps (poles, coops, pickets, etc.) will need to be 12 feet wide. For combined training events or unrecognized shows, it may be 10 or 12 feet wide.

If you need jumps in the 3 or 4 foot range, consider using gates, panels and pickets in heights of 2, 2 1/2, or 3 feet.

When determining the height, always plan on using a pole over the top of the obstacles, which will add 6 inches to the total height. The pole will protect the jumps from the many raps they will take over time. After putting in time to build and paint the jumps, you won't be in any hurry to see them slowly disintegrate!

You should always hang panel, planks, gates and pickets in flat jump cups. If the horse makes a jumping mistake, the pole or obstacle will merely slide out of the cup, thus protecting the jump, the horse and the rider. Since the "handles" (the end part of the jump that hangs in the cups) of these obstacles aren't round and are heavy, they don't dislodge from the regular jump cups unless hit quite hard.

Say, for example, you plan to offer hunter classes with the height of the fences set at 2 feet for ponies and 3 feet for horses. The budget dictates that you can only afford one of each obstacle so you want to use the same obstacles for both ponies and horses. In that case, gates, panels, coops, etc. should be 18 inches high. When the pony classes are over, the jump can be raised to 3 feet by adding poles, raising the gate or panel, etc., and using ground lines to fill the gap between the ground and the jump. A pole or flower boxes on the ground in front of the jump serve well as ground lines.

If you need jumps in the 3 or 4 foot range, consider using gates, panels and pickets in heights of 2, 2½ or 3 feet. The pole on top will add 6 inches to the total height. Though the standards have holes every 3 inches, the diameter of the poles will be 3½ to 4 inches.

Cut evergreens, flower boxes or brush boxes can also add dimension to the jumps' ground lines for variety, color and to make the horses jump rounder. The same points also apply to jumper and combined training classes offering different heights.

For AHSA recognized hunter classes, wing standards must

be used along with 12-foot wide jumps. Regarding the width of obstacles in hunter classes, the AHSA rule book states, "Every course must have at least four different type obstacles. All obstacles must be at least 20' wide or have wings at least 30" wide that are at least 12" higher than the obstacle."

The AHSA jumper classes rule regarding the width of an obstacle simply states, "Rails must be at least six feet long."

The USCTA rules for show jump obstacles are not as precise regarding a minimum width. The rules that govern the height and spread of the obstacles are the same for the show jump phase as the cross country phase.

Jumps For Schooling Only

For schooling beginner jumping riders or for starting young horses over fences, the instructor or trainer may only need a basic few jumps. Four to six pairs of schooling standards with 16- to 20-foot poles should do nicely. With these basics, ground poles can be laid out for trotting over to develop rhythm and to muscle-up the young horse, or to teach balance in the two point position to the rider.

As the horse or rider progresses, gymnastic jumping grids can be set up for two, and then one stride combinations. Vertical fences and oxers can be erected to provide variety.

If training beyond the basics is required, other obstacles can be built to introduce the horse or rider to various types of jumps. A coop, though cumbersome to move, prepares the horse and rider for both the show ring and the hunt field. Gates and pickets provide variety and are simpler to make. Panels and planks offer endless possibilities for those who like to paint, perhaps a basic stone or brick wall on one side, and then paint creative designs for the other side.

A set of cavalletti provides excellent schooling for the green horse and rider to build confidence, rhythm and balance. For the horse especially, trotting over cavalletti is analogous to arm curls with 5 to 10 pound weights for the human athlete.

Cavalletti are safer than ground poles because they do not roll when bumped by the horse, but stay in place. Many a horse has strained a leg stepping on a ground pole that subsequently rolled under the foot.

Classical dressage trainers use cavalletti not only to build rhythm and balance, but also to strengthen the muscles of the horse's back, loins, hindquarters and abdomen. Several classic books on training describe the proper use of cavalletti and some are listed for you in the appendix.

Sturdy Versus Lightweight Jumps

For some, it's tempting to build jumps that are lightweight. The reasons vary from: they are easier to move around; they won't hurt the horse; and the materials are cheaper. But two important facts remain: the sturdier the jump is, the longer it will last and the more the horse will respect it.

One way to make a jump lighter for schooling is to make all obstacles and poles 10-foot wide. Two feet less of lumber makes the jumps a bit easier to move around for children and small women. People with back ailments should not move jumps at all. Even with poles, the two foot difference makes the 12-footers a bit more awkward to maneuver and balance than the 10-footers. With other obstacles like panels, gates, and pickets, the cumulative extra lumber per foot really compounds the weight.

The question of whether lighter jumps "won't hurt the horse" is highly debatable. Take the case of wood versus PVC jump poles. Some people have gone to using PVC because it "won't hurt the horse" and also because finding milled 10 or 12 foot poles at a lumber yard is very difficult if not unlikely. PVC poles are about the same price or a little more expensive than 4x4x10 or 4x4x12 poles.

What happens with PVC, though, is two-fold problem. Yes, the poles are significantly lighter and easier to move around, but the horse also finds them lighter too. The lazy or insensitive jumper may find it easier to brush or rub the pole than to make an honest jumping effort. The lazy jumper then becomes an unsafe jumper when confronting solid jumps in combined training or in the hunt field. Also, the horse that frequently knocks rails down will not be very desirable for the show ring either. And careless jumpers are difficult to sell! It's a very bad habit for sure.

Most trainers prefer jump poles that are heavy enough to give the horse a good sting if they rap it going over the fence, but will also come down if hit too hard. If the pole stings, the next time the horse will more likely clear it. A clean jumper makes a safe and honest jumper, too. It's safe for the horse's legs and safer for the rider.

Using lumber of thicker dimensions will enable the obstacles to stand up to jumping mistakes better too. Take, for example, the plywood recommended on the panel. We always use ⅝"-thick outdoor quality plywood. Yes, it's heavy stuff. On certain days, I can carry a 10 ft. x 2 ft. panel, but on other days I have to drag it. But if it's between ¼ or ⅝ inch plywood, which one do you think will hold up best when a horse stops awkwardly and plants a shod hoof in the center of it? We've all seen panels set at an angle somewhere with the bottom center of the plywood punched out in the size of a hoof.

Many obstacles like the picket, panel and gate are awkward to move anyway no matter what the weight. Dragging them around the arena, they will always want to turn down in your hands if you're holding it just by the handle, the part of the jump that rests in the jump cups. It's always easier to work with a partner when setting up a course.

Remember the adage to always bend at the knees and keep your back straight when lifting jumps. A back injury is no fun, plus it may limit your riding.

Chapter 2

Shopping For Materials

Making a shopping list for materials is a matter of pre-planning. All the pieces of lumber used in the jumps and other equipment described in Chapter 4 are industry standard pieces, therefore custom cut lumber is not needed. The lumber will be found at most local home improvement stores or lumber yards.

To make your list, review the plans of the jump you plan to construct and write down the exact lumber pieces needed. It may also help to jot down what the lumber is for in case that piece is not available in the store or lumber yard. Sometimes you may have to make a substitution and purchase a longer piece than planned for. Or instead of a 12 foot piece, you may find that you can buy two 6 foot pieces. But be sure you know what each piece is for before making a compensation purchase. It's a nuisance to return lumber.

It may be a good idea to call the store or lumber yard before you go to see if they normally stock what you need and if they currently have the lumber you need available. If they don't have what you need, check with another store if one is available. For example, one store may stock 1x6x12's, but not 1x6x10's. With a quick phone call, you may find that another store in town will have 1x6x10's which will be quite handy if you were planning to make a 10 foot wide gate. If the 1x6x10 is unavailable, then plan on buying the 12 foot piece and cutting off 2 feet.

If you are making many different types of jumps, you may find that you need to go to more than one store to get all the lumber on the list. This is especially true during the heavy deck building season when stores frequently deplete their stock to contractors. On the other hand, in winter the stores may not stock the full line of lumber that they did in the warmer months due to the slowdown in outdoor construction. Deck builders are the primary users of the pressure-treated lumber which you will be using on most of the jumps.

The nails called for in the plans are a normally stocked item in most home improvement stores or lumber yards. The best nails to buy for outdoor use are galvanized nails. They don't rust like bright (shiny) nails, but do cost a little bit more. It will be worth the price because bright nails will rust after a season outdoors and rain will make streaked rust marks. Even when painted over, bright nails can rust. Under white paint, the orange rust streaks will show right through, marring the appearance of your hard work.

Use galvanized spiral decking nails if available. These hold the lumber pieces together more tightly than smooth common nails.

Ribbed-type deck nails are only recommended for those who can swing a hammer with accuracy. They hold the tightest and the best—almost—a permanent hold. And here's a word of caution: if you make a mistake nailing one in, either by nailing it crooked or in the wrong spot, it is very difficult to pull the nail back out again. In fact, they hold so well you will probably tear up the pieces of lumber they were nailed into before you can get it out. That can mean, you have to go back to the store and buy another piece of lumber! So spiral nails are strongly recommended instead of ribbed nails.

Three basic types of galvanized nails are the common nail, the spiral decking nail and the ribbed decking nail. We use the spiral nail for a good hold when building jumps.

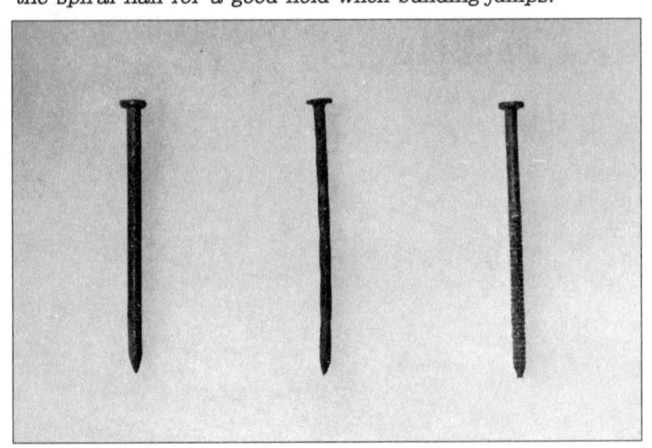

Should you find yourself needing to remove deck nails in order to start again, use a crow bar to remove them. Crow bars are designed to give more leverage to help pry nails out of wood. If necessary, you may need to slip the edge of the crow bar between the pieces of wood to get started.

Pressure-Treated Lumber Versus Untreated Lumber

Pressure-treated lumber is pine that has had water and chromated copper arsenate forced into it to discourage bugs from chewing on the wood when the wood is in contact with the ground. While pressure treating does extend the life of the wood, considering that pine is among the softest woods, don't count on their usual 40 year guarantee.

Nevertheless, pressure-treated pine is quite suitable for

making jumps. It is by far cheaper and more readily available than hardwoods. Hardwoods are also considerably heavier.

Pressure-treated lumber comes in two grades. Grade one had been chemically treated and then kiln-dried. The advantage of kiln-dried lumber is that it is dry, has less knots and is usually straight. Because it is kiln-dried, it is also more expensive and not many home improvement stores stock it.

Grade two lumber is cheaper and much more likely to be stocked in your local home improvement store and lumber yard. When new bundles of grade two lumber are delivered to the stores, they are usually still damp through and through from the pressure-treating process. When damp, the wood is usually still straight.

Watch Out for Warped Wood

Guarding against warped lumber saves a lot of frustration during all phases of jump building. Prevention begins at the lumber yard by purchasing only straight lumber. Handle each piece of lumber when selecting the wood needed.

Pick up one end and sight down the entire length, looking for deviations in the straight smooth surface. Defects are many, and may include pieces that have a cup shape to the surface, a crook that may serve better as a rocking chair base, a bow or a twist.

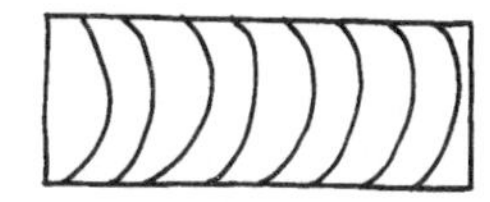

Lumber with annual rings that run parallel to the narrow edge are warp resistant. Those that have annual rings parallel to the wide face have a tendency to warp.

If the piece appears straight, next look at the annual rings visible at ends of the piece. Lumber with annual rings that run parallel to the narrow edge are warp resistant. Those that have annual rings parallel to the wide face of the board have a tendency to warp with changes in humidity and temperature. This is critical with pressure-treated wood because it will begin to dry out once you bring it home.

After you bring the lumber home, if it is stored outside in warm temperatures it will begin to dry out. As it dries, caution must be used to prevent the lumber from warping. The wood begins to twist and warp when the wood dries unevenly.

To prevent warp damage to your newly purchased lumber, there are a few things you can do.

- ✔ Use it immediately. Once it is cut up and nailed together, the pieces will dry as they are formed.
- ✔ If you can't get to it immediately, store the lumber off of the ground on saw horses, on blocks, or in some fashion to allow air circulation underneath it. Store it where the sun won't bake it.

Poles (4x4x10's and 4x4x12's) will be your biggest challenge. The slower they dry, the better. After they are cut and in use, never leave them laying on the ground any longer than 12 hours until they are completely cured. After they have cured (completely dry with cracks showing), if you don't plan to use them for a while, store them with at least one end off the ground. When poles lay on the ground for prolonged periods, they may take on ground moisture and warp, especially if not painted.

As pressure-treated lumber dries, the wood will begin to crack longitudinally. This is normal. Water and chemicals were forced into the wood fibers as part of the pressure treating process. The cracked wood still retains its strength because the long wood fibers are still intact. Also, as it dries, the wood will become lighter. After the wood is cured, you may find some pieces, like poles, are lighter than others. This is because some tree rings are denser than others.

If you are planning to paint your jump equipment, wait until the cracks show. At that point the wood is dry enough for the paint to bond to the wood. (See Chapter 5 for more information on painting pressure-treated lumber.)

Untreated wood can also be used in your jump building projects. It comes in most of the sizes called for in the jump plans except for the 4x4's. These are usually only available in treated wood at home improvement stores. Unless you plan to paint the jumps and keep them painted, untreated wood will not last as long as treated wood.

When making jumps that you definitely plan to paint like panels or planks, buying untreated plywood and top pieces for the panel and 2x12x12's for the planks is recommended. Paint bonds much more successfully to untreated wood and your specialized painting will last much longer.

Another point of interest with lumber is the difference between how it is labeled and its actual size. For kiln dried lumber, a 2x4 is actually 1½ inches by 3½ inches, and for treated wood it's 1¾ inches by 3¾ inches. This is because, historically, all lumber sizes are based on the dimension of green lumber as it comes from the sawmill.

The difference between how lumber is labeled and the actual size is an important consideration in measurements. When marking the center along a 4x4 for drilling jump cup holes, the center will not be at 2 inches but actually at about 1½ to 1¾ inches.

Chapter 3

The Tools You'll Need

The tools required for each item in Chapter 4 are listed on the shopping list. For nearly every job in this book, the builder can successfully make jumps with:

- ✔ a claw hammer
- ✔ a circular saw
- ✔ an electric drill
- ✔ 2 wood clamps
- ✔ a 25 ft. tape measure
- ✔ pencils

The claw hammer is the one most people have in their homes, even if living in an apartment. For most uses, one with a wooden handle and a steel head is the best and not so heavy to swing as an all steel hammer.

If you purchase a power saw for the first time in order to make jumps, choose a portable circular saw. A 7¼ in., 2¾ horsepower circular saw fits the bill nicely for sawing heavy treated lumber. Not only can it be used to make jumps, it is quite useful

The basic tools needed to build jumps are a circular saw, and electric drill, a claw hammer, and a 25 ft. tape measure.

when making barn repairs and replacing boards on the paddock fence.

A circular saw is more economical and versatile as opposed to a table saw or a radial arm saw. Most lumber pieces, especially the 4x4's and the other 10 and 12 foot pieces, are very heavy and awkward to maneuver. In fact, the wood is usually heavier than the circular saw itself. The work is easier and safer if the lumber is clamped or anchored on a work table or a pair of saw horses.

While a table saw can be used to make cross cuts or rip cuts (cuts along the length of the lumber) for smaller pieces of wood, maneuvering a 20 to 25 pound damp piece of pressure-treated lumber steadily across a table saw to make a precise cut is very difficult.

Since pressure-treated wood is the most common for making jumps, a special rip-cut saw blade with a teflon (or other coating) is highly recommended. It will be labeled as a "deck building" blade in the home improvement store. It cuts cleanly through the damp pressure-treated wood best without binding.

Saw blades heat up from the friction during use and the teflon coating prevents chemicals and pine sap from building up a film on the blade. On uncoated blades, as the film becomes thicker, it creates additional friction and slows the blade down during cutting, putting an extra load on the saw's motor. Eventually the saw motor will burn out with such strenuous use.

Have pencils on hand for marking and drawing lines. These don't need to be extra sharp, just available.

An electric drill is a necessity for drilling holes for the jump cups in standards, through poles and the cross legs of cavalletti, and for the bases of the wing standards. Using a manual drill on pressure-treated 4x4's would be near impossible. A light weight electric drill may get the job done for a few standards, but may burn out after drilling several damp pieces of treated wood.

If buying a drill for the first time for making jumps, buy the best that your budget will allow. You can rarely go wrong buying too much power for other uses.

For best results, buy one that has a ½ inch chuck, variable speeds and is reversible. (The chuck is the clamping device on the drill for holding the drill bit.) The drill bit will jam in the 4x4's occasionally and reversing the direction of the drill will allow the bit to be extracted with ease.

Wood clamps are indispensable for anchoring lumber in order to cut with accuracy and safety. Fasten the lumber securely to a work table or saw horses and the lumber will remain stationary. Clamping the wood and using two hands for running a circular saw will make for a straight cut and safe working conditions. Unsecured lumber will move as it is cut, making for a sloppy cut. It may even become a projectile, launched by the spinning blade.

A 25 in. tape measure is essential for measuring. All measurements must be as precise as possible during all phases of construction. Several small measuring errors will eventually compound into a big problem towards the completion of a project. Have pencils on hand for marking and drawing lines. These don't need to be extra sharp, just available.

Additional Tools To Make The Work Easier and More Accurate

Like with anything else, having the right tool for the right job makes it a lot easier. Additional items that are relatively inexpensive but can help a lot include:

- ✔ a carpenter's triangle with 90 degree/45 degree edges, an aid to make straight pencil lines on lumber for cutting.
- ✔ a framing square, an aid to make jumps square and true.
- ✔ a #3 phillips head drill bit. This makes sinking those screws easier.
- ✔ a jigsaw, for cutting shapes and in tight areas.

A carpenter's triangle has ruler markings on it, and a lip to grip the edge of the lumber for marking lines and precise measurements. It has the correct angle for 45 degree lines. With it you can mark straight lines for cross cuts and 45 degree cuts on the lumber. It works well in finding and marking the center of the 4x4's for drilling the jump cup holes, too.

A framing square has a 12-inch edge and an 18-inch edge. This L-shaped ruler enables you to mark longer lines that are correctly parallel or perpendicular to an edge. It is useful in determining if panels, gates and pickets are square instead of oblique.

A #3 phillips head drill bit is inexpensive and helps you screw in wood screws quickly. Be sure to start the screw slowly with a steady drill speed, or the drill bit may quickly strip the screw head.

A jigsaw is useful for trimming the inside corners where the 7¼ inch sawblade may leave joints at an angle. It can also be used to cut decorative shapes on signs or other creative jumps.

If you are planning to build a lot of jumps, or even to start a business of building jumps, certain tools are a must. They save time and last longer under heavy use These include, but are not limited to:

Helpful tools for measuring accurately are the carpenter's triangle, a framing square, a tape measure, a pencil and a 12-inch ruler.

A ½ in. electric drill with a ¾ in. chuck works well for drilling with the larger drill bits. The heavier motor will drill through the thick 4x4's much more efficiently than a light weight electrical drill.

For serious drilling, an industrial-rated drill press works with incredible strength, marching the drill bit through the heavy 4x4's with ease. A two horsepower electric motor with a waist high table is just the ticket. The bit is lowered onto the lumber with a rotating handle. The most important aspect of a drill press is the capability of drilling perfectly straight, centered holes through 4x4's, which can be a battle with hand-held drills.

A radial arm saw enables you to make precise and steady cuts on 2 inches or less thick lumber pieces without taking the time to clamp the wood. The 4x4's will still have to be cut a half at a time.

Left: For serious drilling, an industrial rated drill press makes drilling through 4x4's easy and the holes straight.

Below: The jigsaw and electric sander help with trimming corners and smoothing rough edges

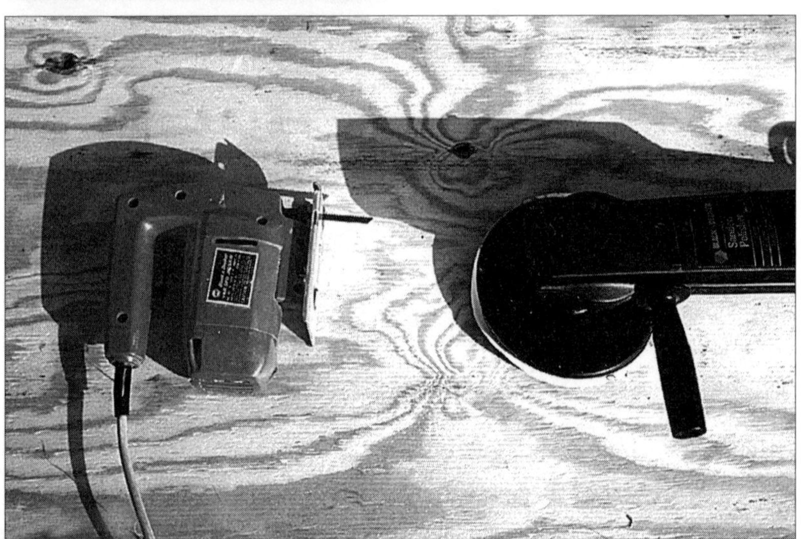

An electric sander is useful for smoothing frayed edges on cut lumber to present a finished looking product.

Personal Safety

All power tools are dangerous when used improperly. When purchasing new power tools or borrowing power tools, always read the information manual cover to cover before using the equipment. The manuals detail the safe handling and use of the power tool with the extra plus that you may learn some things about it that you didn't know it could do for you. If you haven't used your tools in a while, review their booklets.

The most dangerous aspect of using a circular saw is the danger of "kick-back." The manual will explain how to use it safely and prevent this situation.

For personal protection, always wear safety glasses while sawing, drilling and hammering. They protect the eyes from flying sawdust, wood chips, and nails.

When sawing or drilling, especially with treated wood,

always wear a dust mask. The chemicals used in the process are unhealthy to the respiratory system and anyone with allergies will certainly be affected by the fine sawdust produced during sawing and drilling. When possible, do all sawing and drilling outdoors, or in a very well-ventilated area with a fan running.

After handling uncured treated lumber, whether loading or unloading it, or sawing, drilling and hammering it, always wash your hands.

The chromated copper arsenate that is injected into the wood is similar to arsenic, a toxic substance, and can be absorbed into the skin. For loading and unloading the lumber, wear gloves. But do not wear gloves for sawing or drilling lumber. The saw blade or drill bit may take hold of the glove with disastrous results to your hand.

Wear sturdy shoes, like the ones you wear out to the barn. A piece of lumber dropped on the toes is a bit unpleasant! When using power tools, never wear loose clothing. Tuck in shirt tails, roll up loose sleeves, and tie back long hair.

Always paint outdoors if at all possible. Dust masks for painting are available to block most paint fumes. Latex is not as potent as oil base paints, but both produce fumes irritating to the nose, mouth and lungs.

Chapter 4

Shopping Lists and Construction Notes

This chapter contains the shopping lists and construction notes for building various jumps and other stable equipment. All the dimensions in the following drawings and shopping lists are based on 12-foot wide by 2 foot 6 inch jumps. If constructing jumps of 2 feet and below, or 3 feet and above, or in 10-foot widths, make the necessary adjustments in lumber requirements by adding or subtracting the required inches and go from there.

The construction notes assume some basic knowledge of using saws, hammers and drills. There are many useful books at the pubic library and in your local home improvement stores to help explain many simple and complicated tasks in carpentry.

Also, some variations can be applied to the following plans to suit your specific needs. These plans may serve as a springboard for other ideas too.

SHOPPING LIST

Use only pressure-treated lumber.

For two standards:

4 ft. height:

- ✔One 4x4x8 for the posts
- ✔One 2x6x8 for the base
- ✔16d decking nails

5 ft. height:

same as above except use one 4x4x10 cut in half.

6 ft. height:

same as above except use two 4x4x6's.

TOOLS NEEDED

- ✔Circular saw
- ✔Electrical drill
- ✔9/16" or 5/8" wood auger drill bit
- ✔Hammer
- ✔Tape measure
- ✔Carpenter's triangle
- ✔Pencils

Basic Schooling Standards

The basic schooling standard is the easiest standard to make.

Construction Notes

1. Cut the 4x4 in half.
2. Mark and drill the jump cup holes with a 9/16" or 5/8" wood auger drill bit every 3 in. starting the first hole 11 in. from the bottom. Care must be taken to drill the jump cup holes in the center of the post. Drilling a hole off center may mean your jump cups will only fit on the post from one side. While the better made heavy steel cups allow for up to 2 inches from the edge to the hole, some less expensive cups will may not fit as well. The width of 4x4's will vary somewhat, but the carpenter's triangle will enable you to find the center. Make several centered marks about 6 in. apart along the post. Then use a straight edge to connect the lines. Next, make a pencil mark every 3 in. Use a hammer to tap a nail to punch a dent in the wood at each mark. This will enable the drill bit to start exactly where it's supposed to, instead of wandering around and grabbing a start in the wrong location.
3. Cut the 2x6 into lengths as specified in the plans.
4. Nail on each 2x6 as indicated on plans with two 16d nails.

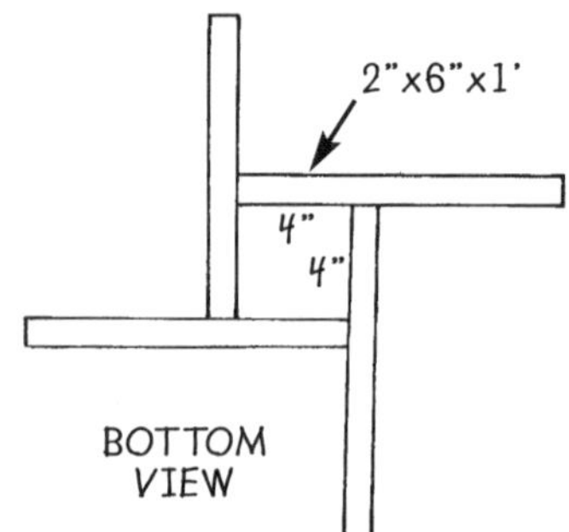

Deluxe Schooling Standards

This close-up shows the base of the deluxe schooling standard. The design seems somewhat complex, but it is extremely durable over many years of hard use.

Construction Notes

1. Cut the 4x4 in half.
2. Mark and drill the jump cup holes with a 9/16" or 5/8" wood auger drill bit every 3 in., starting the first hole at 14 in. from bottom. Care must be taken to drill the jump cup holes in the center of the post. Drilling a hole off center may mean your jump cups will only fit on the post from one side. While the better made heavy steel cups allow for up to 2 inches from the edge to the hole, some less expensive cups may not fit as well. The width of the 4x4's will vary somewhat, but the carpenter's triangle will enable you to find the center. Make several centered marks about 6 in. apart along the post.

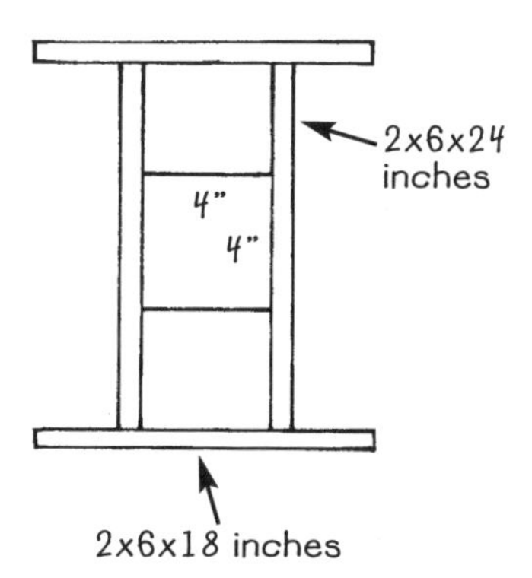

SHOPPING LIST

Use only pressure-treated lumber.

For two standards:

4 ft. height:

- ✔One 4x4x8 for the posts
- ✔One 2x4x8 for the supports
- ✔Two 2x6x8's for the bases
- ✔16d and 8d decking nails

5 ft. height:

same as above except use one 4x4x10 cut in half.

6 ft. height:

same as above except use two 4x4x6's.

TOOLS NEEDED

- ✔Circular saw
- ✔Electrical drill
- ✔9/16" or 5/8" wood auger drill bit
- ✔Hammer
- ✔Tape measure
- ✔Carpenter's triangle
- ✔Pencils

Then use a straight edge to connect the lines. Next, make a pencil mark every 3 in., beginning at 14 in. Use a hammer to tap a nail to punch a dent in the wood at each mark. This will enable the drill bit to start exactly where it's supposed to, instead of wandering around and grabbing a start in the wrong location.

3. Cut the 2x6 into lengths as specified on the plans.
4. Cut two 2x4's at 45 degree angles, with the longest side 14¼ in.
5. Nail the first 2 ft. 2x6x24 on the 4x4 on the side without holes with two 16d nails.
6. Nail on the angled 2x4's with 8d nails.
7. Nail on the other 2 ft. 2x6.
8. Nail the 2x6x18's on the ends.

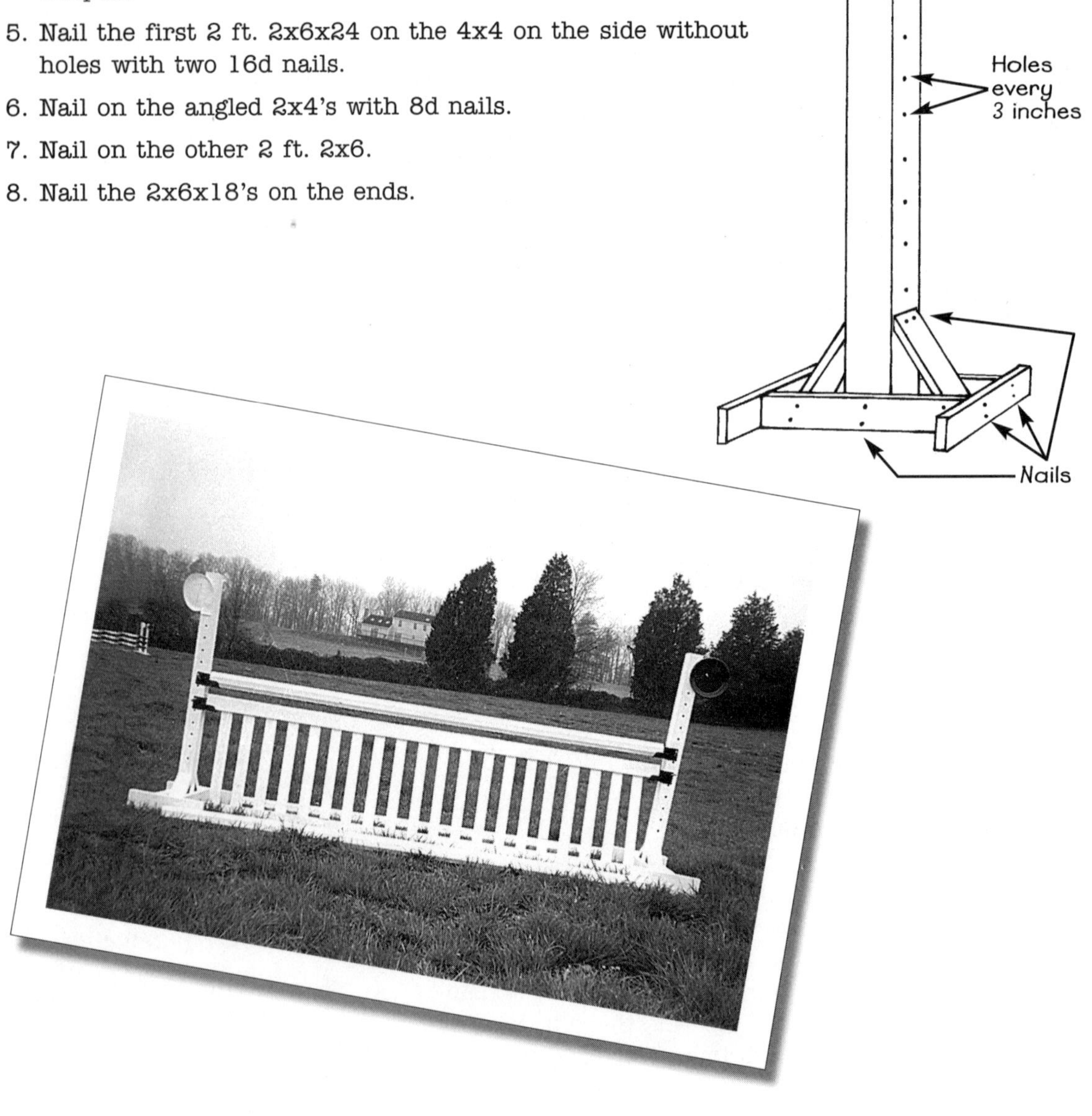

Wing Standards

This shot shows the right side of 5 ft. wing standard.

Construction notes

1. Cut the 4x4's in half.
2. Mark and drill the jump cut holes every 3 inches in the posts starting at 11 inches from bottom. Care must be taken to drill the jump cup holes in the center of the post. Drilling a hole off center may mean your jump cups will only fit on the post from one side. While the better made heavy steel cups allow for up to 2 inches from the edge to the hole, some less expensive cups will may not fit as well. The width of 4x4's will vary somewhat, but the carpenter's triangle will enable you to find the center. Make several centered marks about 6 in. apart along the post. Then use a straight edge to connect the lines. Next, make a pencil mark every 3 in. Use a hammer to tap a nail to punch a dent in the wood at each mark. This will enable the drill bit to start exactly where it's supposed to, instead of wandering around and grabbing a start in the wrong location.

SHOPPING LIST

All lumber should be pressure-treated.

Five ft. height:

- ✔Two 4x4x10's for the posts
- ✔One 2x4x10 for the crosspieces
- ✔Two 1x4x12's or four 1x4x6's for the inner pickets
- ✔One 2x8x12 for the bases
- ✔6d decking nails
- ✔3x10 wood screws with phillips heads
- ✔Eight 1/2 x 6" carriage bolts with flat washers & nuts

Six ft. height:

all the same except use four 4x4x6's

TOOLS NEEDED

- ✔Circular saw
- ✔Electrical drill
- ✔$^9/_{16}$" or $^5/_8$" wood auger drill bit
- ✔#3 head phillips drill bit
- ✔Hammer
- ✔Carpenter's triangle
- ✔Tape measure
- ✔Pencils

3. Turn the post a quarter turn, and mark and drill the holes for the base at 1½ in. and 4 in. from bottom.
4. Cut the 2x4 crosspieces at 30 in.
5. Lay the 4x4's parallel on the floor (a flat concrete surface or as level ground as possible) with the bottom of 4x4's butted up against a wall or board for even ground alignment when they are stood up.
6. Lay the 2x4 crosspieces in place centered over the 4th hole from the bottom and 4th hole from the top. (On the 4x4x6's, place the crosspieces at 4th hole from the bottom and 5th hole from the top.)
7. Pencil mark the placement of the 2x4's on the 4x4's for notching the 4x4's. Knowing the exact width is important when marking a 4x4 for notching. A full 4 inch wide notch will be too wide for making wing standards. Always use the lumber itself to mark the width for notching. Lay the 2x4 across the parallel 4x4's where it will be attached and pencil mark the width of the 2x4 for notching.

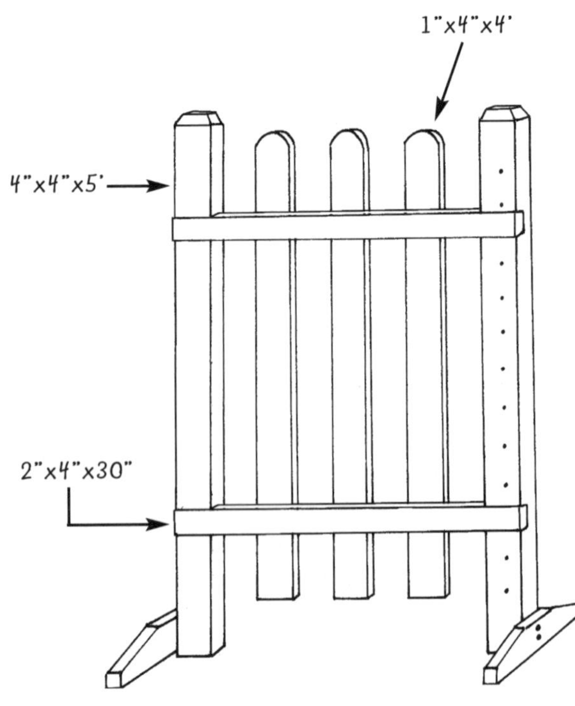

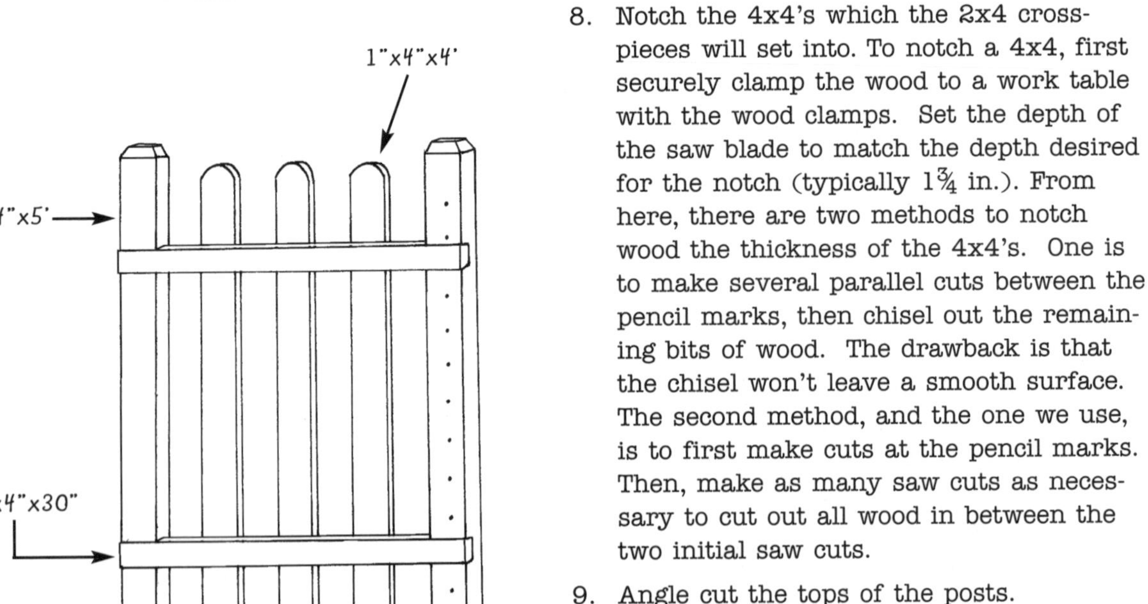

8. Notch the 4x4's which the 2x4 crosspieces will set into. To notch a 4x4, first securely clamp the wood to a work table with the wood clamps. Set the depth of the saw blade to match the depth desired for the notch (typically 1¾ in.). From here, there are two methods to notch wood the thickness of the 4x4's. One is to make several parallel cuts between the pencil marks, then chisel out the remaining bits of wood. The drawback is that the chisel won't leave a smooth surface. The second method, and the one we use, is to first make cuts at the pencil marks. Then, make as many saw cuts as necessary to cut out all wood in between the two initial saw cuts.
9. Angle cut the tops of the posts.
10. Place the 2x4's in the notches, and secure with the 3x10 wood screws using the drill with the #3 drill bit head.

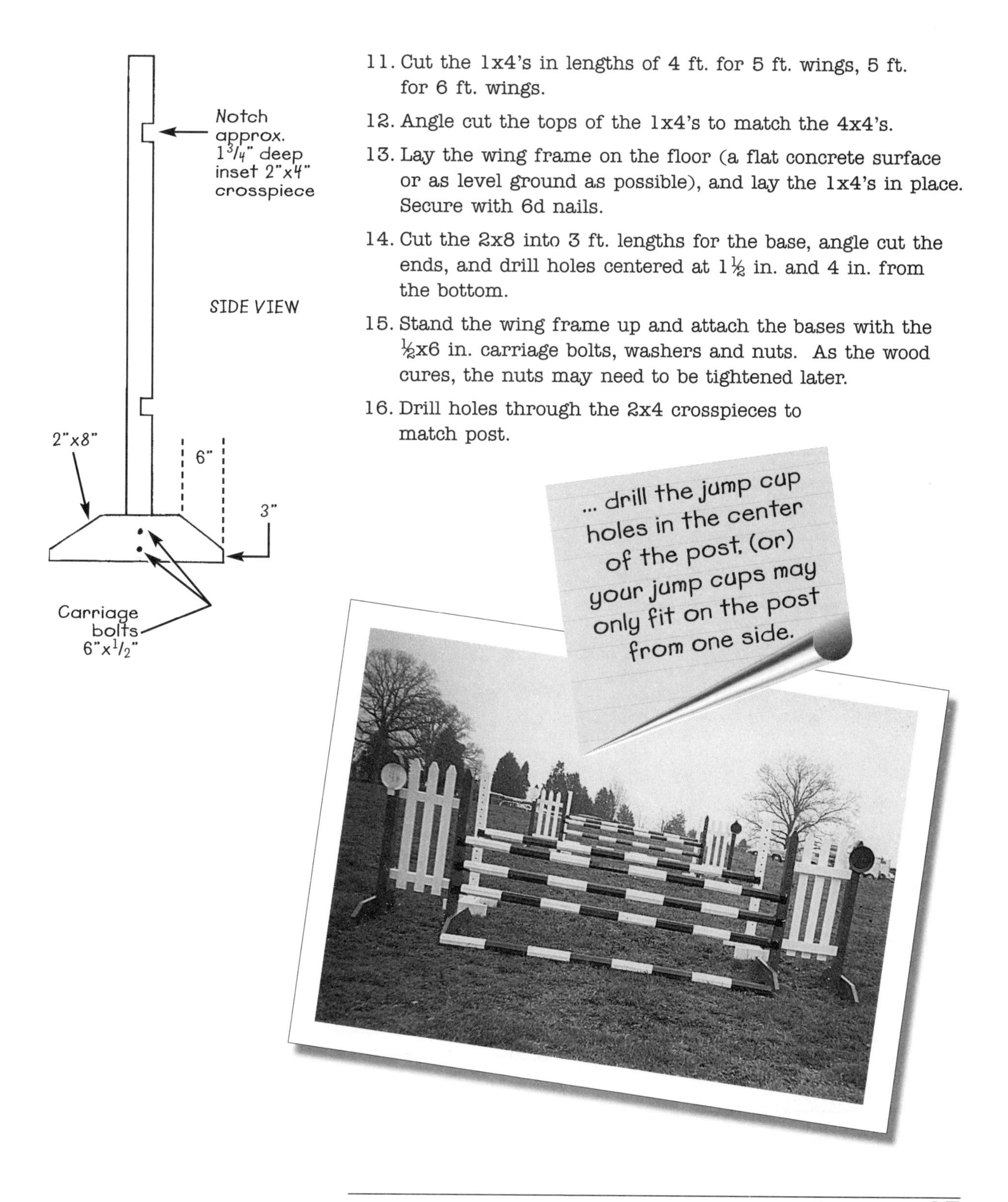

11. Cut the 1x4's in lengths of 4 ft. for 5 ft. wings, 5 ft. for 6 ft. wings.
12. Angle cut the tops of the 1x4's to match the 4x4's.
13. Lay the wing frame on the floor (a flat concrete surface or as level ground as possible), and lay the 1x4's in place. Secure with 6d nails.
14. Cut the 2x8 into 3 ft. lengths for the base, angle cut the ends, and drill holes centered at 1½ in. and 4 in. from the bottom.
15. Stand the wing frame up and attach the bases with the ½x6 in. carriage bolts, washers and nuts. As the wood cures, the nuts may need to be tightened later.
16. Drill holes through the 2x4 crosspieces to match post.

Slant Wing Standards

SHOPPING LIST

Five or Six ft. height.

For two standards:

- ✔One 4x4x10 for the 5 ft. posts
- ✔Two 4x4x6's for the 6 ft. posts
- ✔One 2x4x12 for the cross pieces
- ✔Six 1x4x6's for the inner pickets
- ✔One 2x8x12 for the bases
- ✔6d decking nails
- ✔3x10 wood screws with a phillips head
- ✔Eight 1/2 x 6" carriage bolts with flat washers & nuts

TOOLS NEEDED

- ✔Circular saw
- ✔Electrical drill
- ✔9/16" or 5/8" wood auger drill bit
- ✔#3 head phillips drill bit
- ✔Hammer
- ✔Carpenter's triangle
- ✔Tape measure
- ✔Pencils

This view of the slant standard shows the attachment of the crosspieces and the inner pickets.

Construction Notes

1 Cut the 4x4x10 in half.

2. Mark and drill the jump cup holes every 3 inches in the posts, starting at 11 in. from bottom. Care must be taken to drill the jump cup holes in the center of the post. Drilling a hole off center may mean your jump cups will only fit on the post from one side. While the better made heavy steel cups allow for up to 2 inches from the edge to the hole, some less expensive cups will may not fit as well. The width of the 4x4's will vary somewhat, but the carpenter's triangle will enable you to find the center. Make several centered marks about 6 in. apart along the post. Then use a straight edge to connect the lines. Next, make a pencil mark every 3 in. Use a hammer to tap a nail to punch a dent in the wood at each mark. This will enable the drill bit to start exactly where it's supposed to, instead of wandering around and grabbing a start in the wrong location.

3. Turn the post a quarter turn, and mark and drill the holes for the base at 1½ in. and 4 in. from the bottom.

4. Cut the 2x4 crosspieces at 30 in.

5. Lay the 4x4's parallel on the floor (a flat concrete surface or as level ground as possible), with the bottom butted up against a wall or board for even alignment.
6. Lay the bottom 2x4 crosspiece in place centered over the 4th hole from the bottom.
7. Lay the top crosspiece at an angle over the 4th hole on the 5 ft. 4x4 and the 4th hole on the 6 ft. 4x4.
8. Pencil mark the top 2x4 for angles to cut to match the 4x4's.
9. Pencil mark the placement of both 2x4's on both 4x4's for notching the 4x4's. Knowing the exact width is important when marking a 4x4 for notching. A full 4 inch wide notch will be too wide for making wing standards. Always use the lumber itself to mark the width for notching. Lay the 2x4 across the parallel 4x4's where it will be attached and pencil mark the width of the 2x4 for notching.
10. Notch the 4x4's. To notch a 4x4, first securely clamp the wood to a work table with the wood clamps. Set the depth of the saw blade to match the depth desired for the notch. From here, there are two methods to notch wood the thickness of 4x4's. One is to make several parallel cuts between the pencil marks, then chisel out the remaining bits of wood. The drawback is that the chisel won't leave a smooth surface. The second method, and the one we use, is to first make cuts at the pencil marks. Then, make as many saw cuts as necessary to cut out all wood in between the two initial saw cuts.
11. Angle cut the top of the posts.
12. Place the 2x4's in notches, and secure with 3x10 wood screws with the drill using the #3 phillips head drill bit.

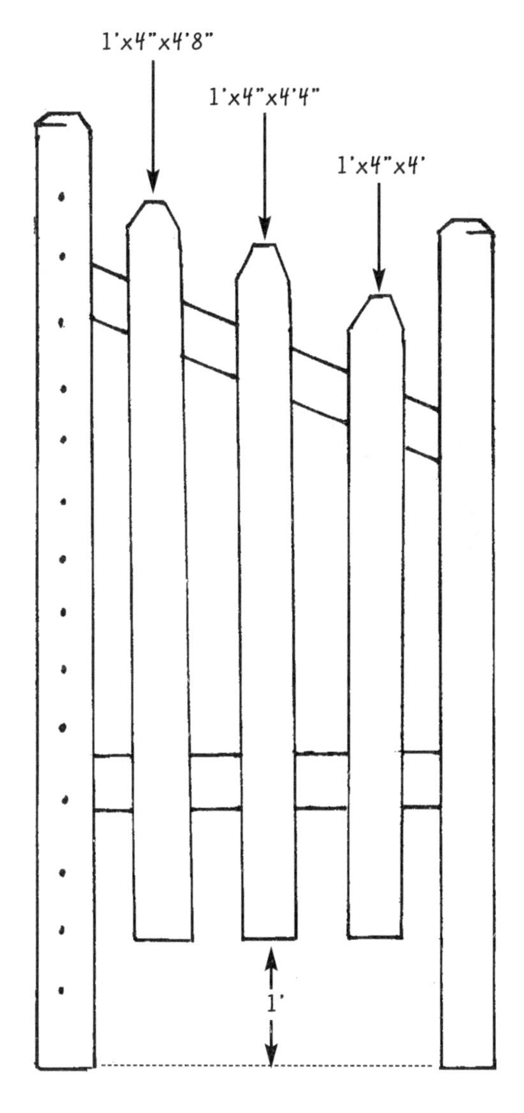

FRONT VIEW

13. Cut the 1x4's in lengths as specified in drawing on page 27.

14 Angle cut the tops of the 1x4's to match the 4x4's.

15. Lay the wing frame on the floor (a flat concrete surface or as level ground as possible), and lay the 1x4's in place. Secure with 6d nails.

16. Cut the 2x8 into 3 ft. lengths, angle cut the ends, and drill holes centered at 1½ in. and 4 in. from the bottom.

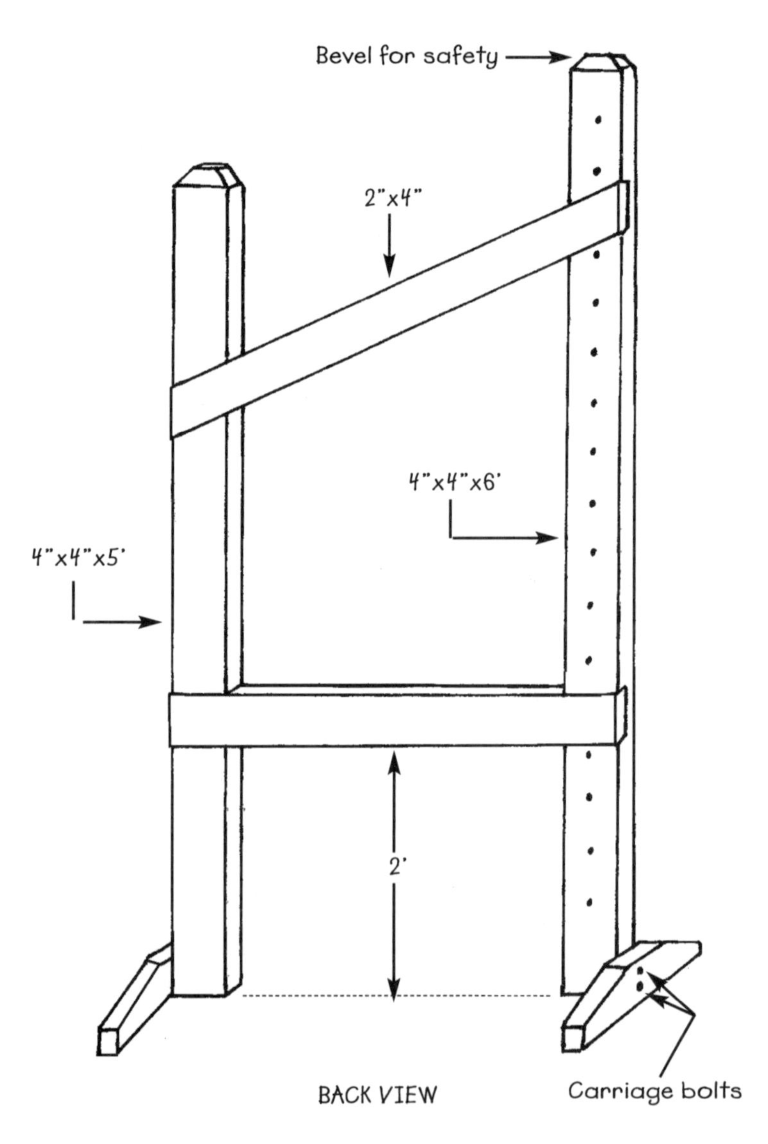

17. Stand the wing frame up and attach the bases with the ½ x 6 in. carriage bolts, washers and nuts. As the wood cures, the nuts may need to be tightened later.

18. Drill holes through 2x4 crosspieces to match the post.

Jump Poles

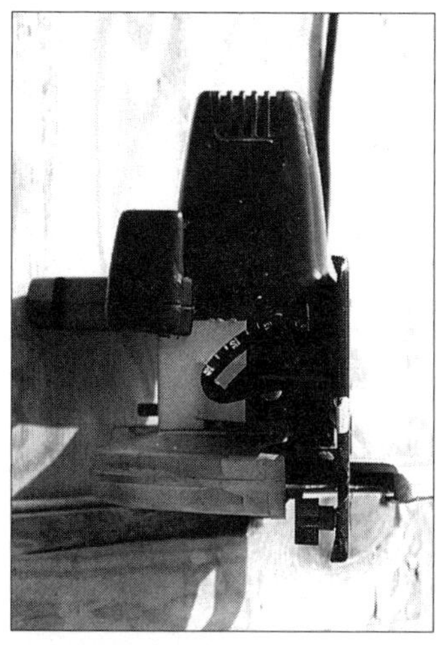

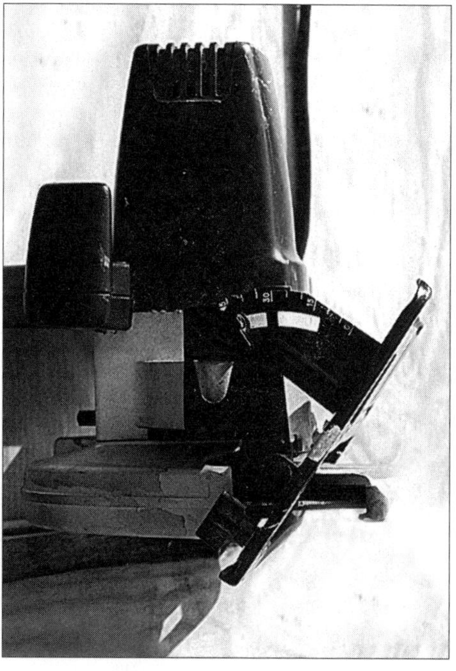

Above: This circular saw is set for normal cuts. the blade angle is set at "0".

Above Right: The blade is adjusted to a 45 degree angle for beveling poles.

Below: A strip of masking tape has been placed 1 in. from the notch on the saw's base to mark the edge of the 4x4 for beveling.

Construction Notes

1. Set up the saw horses on firm, level ground.
2. Set the 4x4's on the saw horses. Secure the 4x4 to be cut with wood clamp so it won't move.
3. Mark a line on the 4x4 1 in. from the edge along the entire length of the 4x4.
4. Set the saw blade at a 45 degree angle.
5. Place a strip of masking tape 1 in. from the notch on the saw's base that guides the blade. The edge of the tape will mark the edge of the 4x4 during the beveling (sawing).
6. Proceed to bevel the edges of the 4x4 by sawing down the entire length of the 4x4.
7. Turn the 4x4 one quarter turn, and repeat step 6 until all four sides have been done.

SHOPPING LIST

All lumber should be pressure-treated.

For 10 ft. poles, buy 4x4x10's.

For 12 ft. poles, buy 4x4x12's.

TOOLS NEEDED

✔Circular saw

✔Two saw horses

✔Two wood clamps

SHOPPING LIST

- ✔Two 1x4x12's untreated wood for the top
- ✔Two 1x4x12's pressure-treated for the bottom
- ✔Two 5/8" outdoor quality plywood sheets (for a 2' high panel, use one plywood sheet)
- ✔6d decking nails
- ✔Two 3d bright finishing nails for tacking on filler pieces

TOOLS NEEDED

- ✔Circular saw
- ✔Hammer
- ✔Two saw horses
- ✔Clamps
- ✔Tape measure
- ✔Carpenter's triangle
- ✔Framing square
- ✔Pencils

12' x 2'6" Panel

This finished panel is 10 ft. x 2 ft. 6 in.

Construction Notes

1. Cut the plywood to the size needed.
2. Cut two 1x4's to the length needed for the bottom of the panel.
3. Cut two filler pieces from the plywood to be placed in between the ends of the top 1x4's for added support to the panel's handles, the part that hangs in the jump cups (see plans).
4. Lay out one top and one bottom 1x4 on the floor (a flat concrete surface or as level ground as possible).
5. Lay the plywood pieces on the 1x4's in place.
6. Tack on the plywood filler pieces with one 3d nail each.
7. Using a tape measure and framing square, verify that the plywood is correctly in place for the right height and correctly squared up to the top 1x4.
8. Lay the other top 1x4 in place on top of panel, and nail on with 6d nails every 5 to 6 in.
9. Lay the other 1x4 in place on the bottom of panel, and nail on with 6d nails every 5 to 6 in.
10. Flip the panel over. Nail in additional 6d nails for a secure hold.

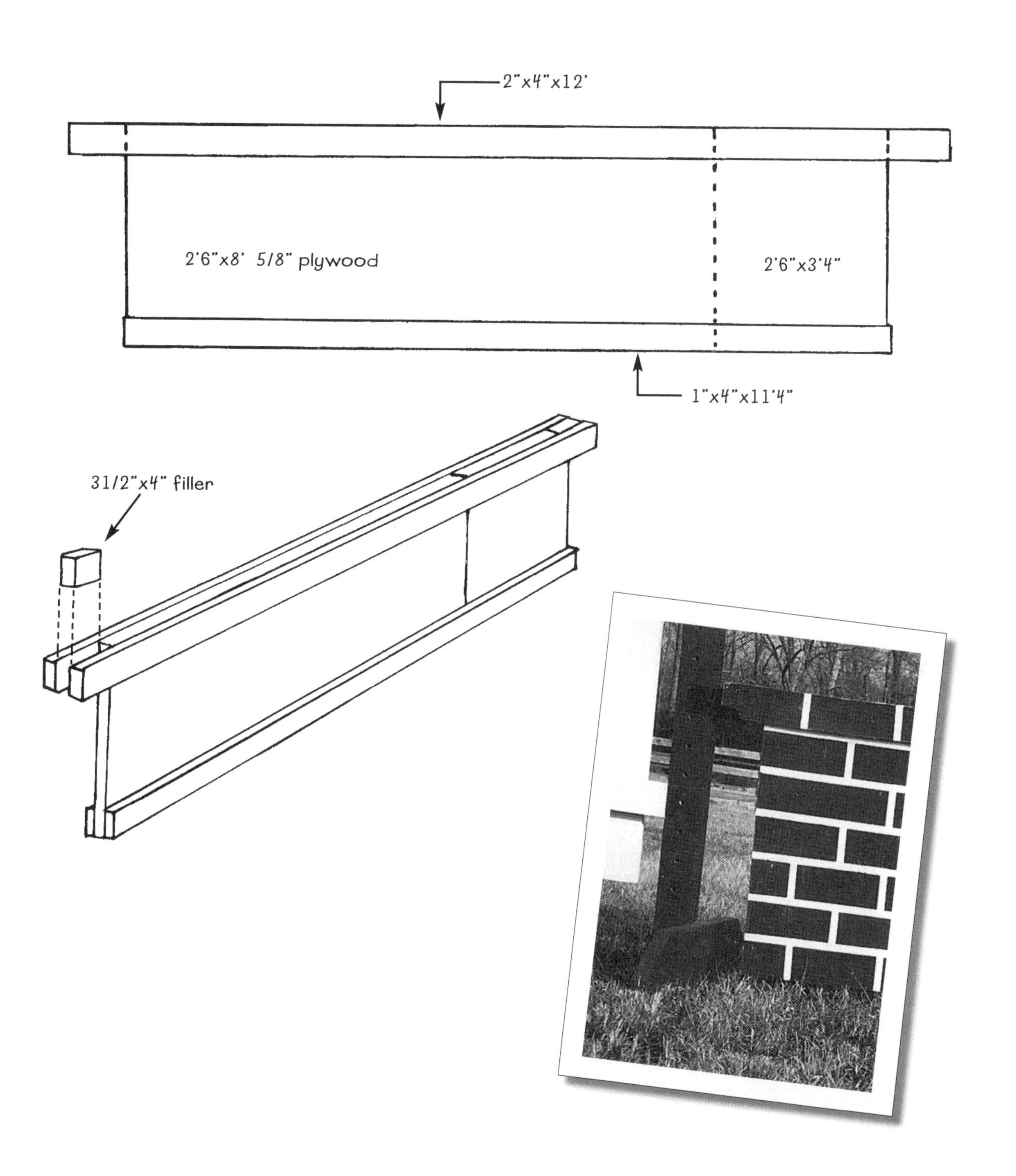
2"x4"x12'
2'6"x8' 5/8" plywood
2'6"x3'4"
1"x4"x11'4"
31/2"x4" filler

SHOPPING LIST

All lumber should be pressure-treated.

- ✔Nine 1x4x12's
- ✔6d decking nails
- ✔Two 3d bright finishing nails for tacking on filler pieces

TOOLS NEEDED:

- ✔Circular saw
- ✔Hammer
- ✔12" ruler
- ✔Tape measure
- ✔Carpenter's triangle
- ✔Framing square
- ✔Pencils

12' x 2'6" Picket

This 10 ft. x 2 ft. 6 in picket is hung in deluxe schooling standards with beveled poles

Construction Notes

1. Cut all 1x4's as indicated on the plans. Cut the two bottom 1x4's to length and the two filler pieces. Cut 20 pickets for a 12 ft.-wide jump, or 18 pickets for a 10 ft.-wide jump.
2. Lay out one 1x4x12 on the floor for the top (a flat concrete surface or as level ground as possible), and one 1x4x11'4" on the floor parallel to the top piece for the bottom.
3. Nail the filler pieces on the ends of the top 1x4 with one 3d finishing nail in the center of each filler.
4. Lay both end pickets in place, aligned against the top filler pieces and squared with the ends of the bottom 1x4.
5. Verify that the top and bottom 1x4's are squared and parallel using a framing square and a tape measure.
6. Measure and pencil mark even spacing for the remaining pickets. Mark the spaces approximately every 7 in. Each space includes both the picket and an open space. This allows for variance in the width of each picket board.

7. Lay the top 1x4 in place.
8. Nail on the top 1x4 with 6d nails through the ends, the fillers, and using one nail for each picket. Verify that each picket is in its proper place before nailing. The vibration of hammering will displace some pickets.
9. Repeat for the bottom 1x4.
10. Flip the picket over. Nail in additional 6d's every other picket and at the ends.

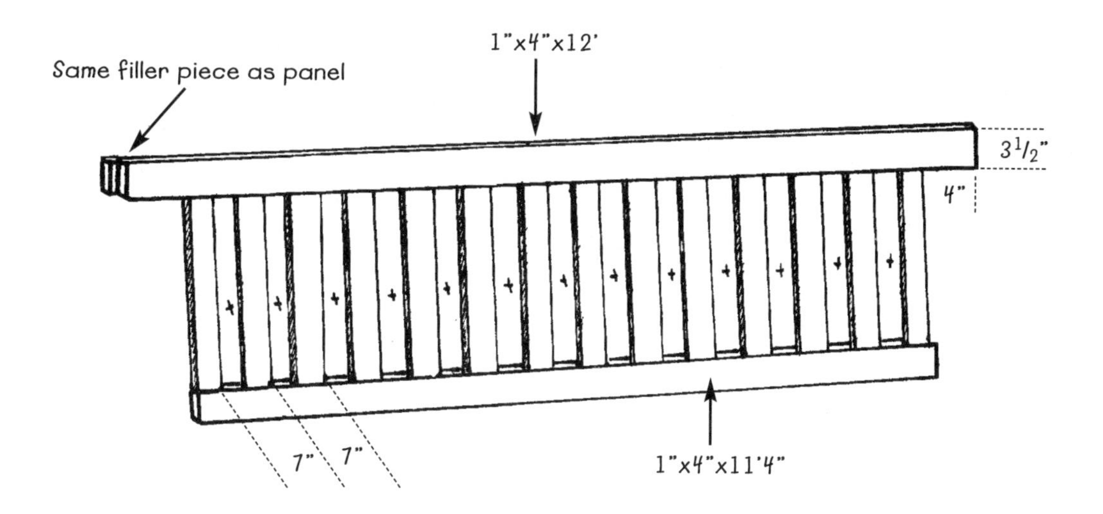

12' x 2'6" Gate

This gate is 12 ft. x 2 ft.

SHOPPING LIST

All lumber should be pressure-treated.

- ✔Seven 1x6x12's
- ✔6d decking nails
- ✔Two 3d bright finishing nails for tacking on filler piece.

TOOLS NEEDED

- ✔Circular saw
- ✔Hammer
- ✔Carpenter's triangle
- ✔Framing square
- ✔Tape measure
- ✔Pencils

Construction Notes

1. Cut the top pieces, bottom pieces, three vertical pieces, and fillers as shown in the drawing.
2. Lay out one top 1x6, and tack on the filler pieces with one 3d bright finishing nail for each piece.
3. Lay out one bottom 1x6 parallel to the top.
4. Set the two end vertical 1x6's in place with the tops aligned against the fillers, and the bottom corners flush with the ends of bottom board.
5. Using the framing square and tape measure, square up the pieces and set to the exact height required (2'6" in this case).
6. Tack the vertical pieces in place with 3d nails.
7. Tack the center board in place with 3d nails.
8. Cut the two angle pieces to 6 ft. length.
9. To mark the angle cuts, lay the 1x6's in place diagonally on the frame, pencil mark either end of cuts to saw. This will be approximately 5 ft. 6 in.
10. Complete the pencil lines and saw the 1x6's.
11. Lay the angle cut boards in place.

12. Lay the second top board in place, nail with 6d nails through the ends, filler pieces, and three vertical boards. Use two nails for each.
13. Lay the second bottom board in place, and nail with 6d nails.
14. Flip the gate over and nail in additional 6d's for a secure hold.

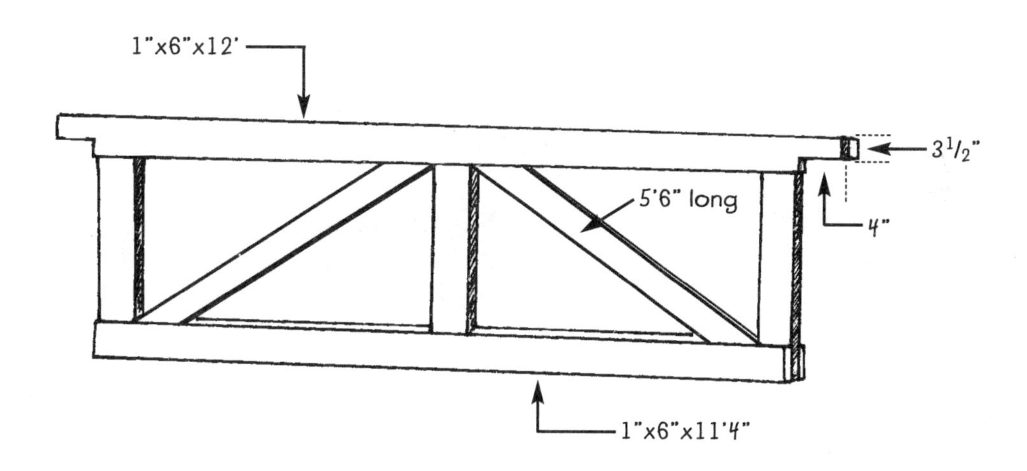

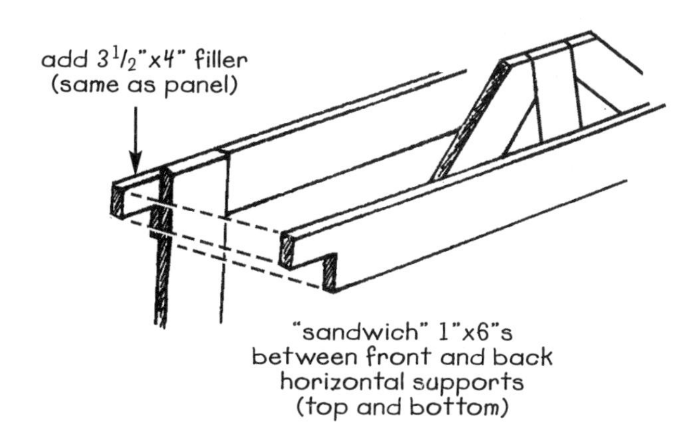

Coop

The coop, finished and painted

SHOPPING LIST

All lumber should be pressure-treated.

Materials needed for one piece coop 10' x 2'6":

- ✔Two 2x4x8's for the frame
- ✔Thirteen 1x6x12's for the outer shell
- ✔8d and 6d decking nails

Materials needed for 12' wide coop in two pieces of 5'9" x 2'6":

- ✔Four 2x4x8's for the frame
- ✔Fourteen 1x6x12's for the outer shell
- ✔One 1x4x12 for the top of the coop
- ✔6d and 16d decking nails

Construction Notes

1. Cut the 2x4's to length.
2. Angle cut the tops of the 2x4's for the top of the frame.
3. Cut the 1x6's to make the crosspieces for frame.
4. Nail the crosspieces on the 2x4's with the 6d's to make the frame.
5. Nail the tops of the frame together with the 6d nails.
6. Cut all the 1x6's for the sides to length.
7. Rip cut (cut the board longitudinally) in half one 1x6 for the top of the coop. If desired, a 1x4 can be used for a wider top.
8. Nail the two bottom 1x6's to the outer surface of the frames with the 8d nails.
9. Nail on the rip cut 1x3 (or 1x4) to the top of coop frame with the 6d's.
10. Nail on the two top 1x6's side boards with the 8d's.
11. Nail on the remaining 1x6's.

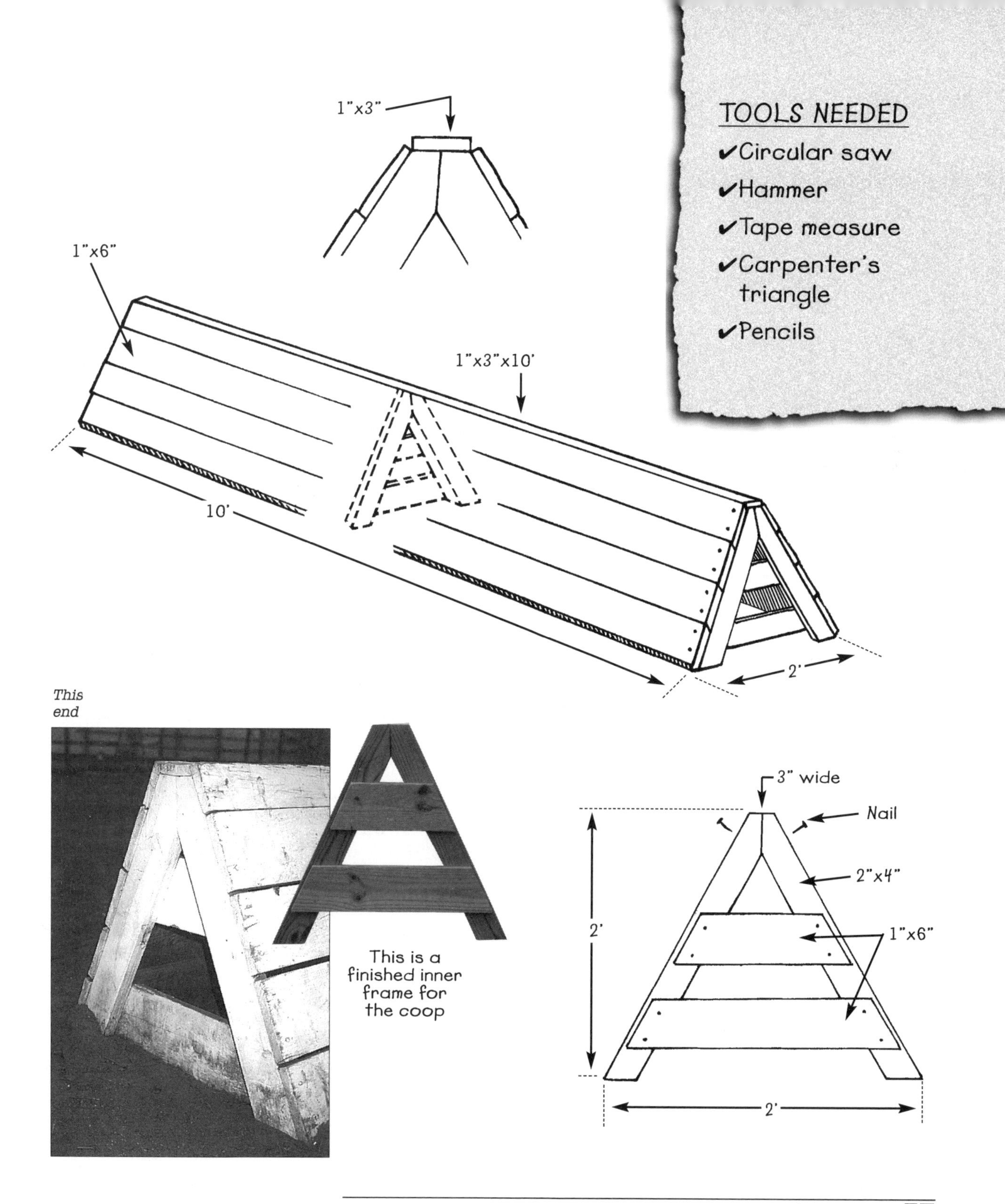

This is a finished inner frame for the coop

TOOLS NEEDED

- ✔Circular saw
- ✔Hammer
- ✔Tape measure
- ✔Carpenter's triangle
- ✔Pencils

8' Cavalletti

This cavalletti is made with a 4 in. diameter fence post.

Construction Notes

1. Mark on both ends of the 4x4x8 the center point at $1\frac{3}{4}$ in. for later drilling.
2. Bevel the edges of the 4x4x8 (see construction notes given earlier for poles). This step is not necessary if using a 3 in. to 4 in. fence post.
3. Cut the 4x4x10 into four 27 in. pieces for the legs.
4. Mark each leg for the width for notching. Knowing the exact width is important when marking a 4x4 for notching. A full 4 inch wide notch will be too wide for making cavalletti. Always use the lumber itself to mark the width for notching. Lay the 4x4 across the other 4x4 where it will be sunk in and pencil mark the width of the 4x4 for notching.
5. Notch the 4x4's. To notch a 4x4, first securely clamp the wood to a work table with the wood clamps. Set the depth of the saw blade to match the depth desired for the notch, typically $1\frac{3}{4}$", which equals half the width of the other 4x4. (A table saw with a dado blade can be used instead of a circular saw for notching.) There are two

SHOPPING LIST

All lumber should be pressure-treated.

- ✔One 4x4x10 for the legs
- ✔One 4x4x8, or a 3 in. to 4 in. x 8 ft. fence post, for the pole
- ✔Two 1/2x10" carriage bolts with flat washers and nuts

TOOLS NEEDED

- ✔Circular saw
- ✔Two saw horses
- ✔Electrical drill
- ✔9/16" wood auger drill bit
- ✔1 1/2" boring drill bit
- ✔3/4" socket for ratchet or 3/4" crescent wrench
- ✔Clamp
- ✔Tape measure
- ✔Carpenter's triangle
- ✔Pencils

methods to notch wood the thickness of the 4x4's. One is to make several parallel cuts between the pencil marks, then chisel out the remaining bits of wood. The drawback is that the chisel won't leave a smooth surface. The second method, and the one we use, is to first make cuts at the pencil marks. Then, make as many saw cuts as necessary to cut out all wood in between the two initial saw cuts.

This detail shows the attachment of the post to the legs. A hole has been bored deep enough for the washer and nut to set below the surface of the pole.

6. Fit the legs together, stand them up and clamp together tightly at the center from the side.
7. Drill through the center of the joint with 9/16" wood auger drill bit.
8. On the beveled 4x4 or fence post, drill with 1½" boring bit to approximately ½ in. depth on the pencil mark from step 1. When the carriage bolt is attached to the leg and pole, the shallow bore hole will allow you to sink the nut and washer below the surface of the pole, leaving no hardware protruding at the top of the pole.
9. Drill through the center of the bore hole with 9/16" wood auger drill bit.
10. Set the beveled 4x4x8 (or fence post) in place in the joint of legs with the holes lined up.
11. Insert the ½ x 10" carriage bolt. Using a ratchet or crescent wrench, secure with washers and nuts. As the wood cures, the nuts may need to be tightened later.

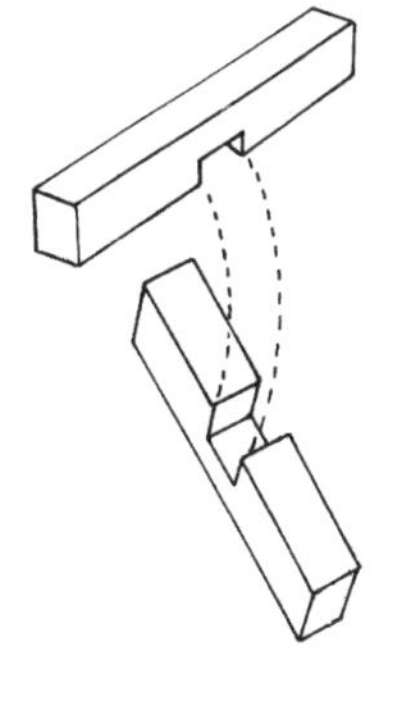

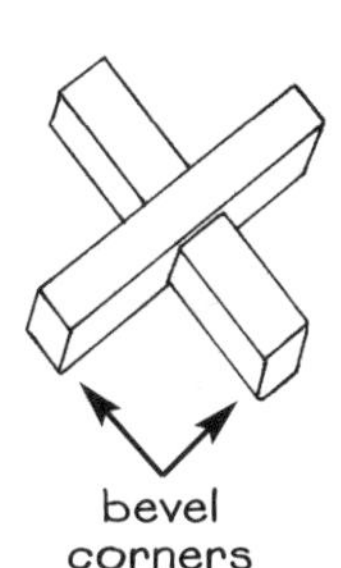

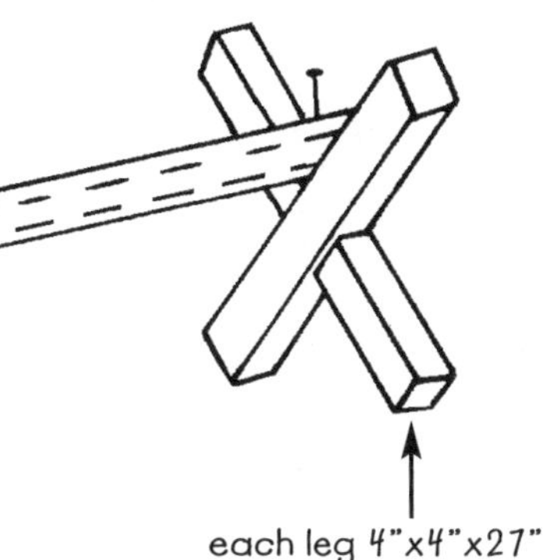

Rolltop

The finished roll top is ready for delivery. Note the grass carpet covering.

Construction Notes

1. Cut the 2x6 and 1x4 pieces to length.
2. Lay the pieces on the floor (a flat concrete surface or as level ground as possible), against each other as shown in the drawing.
3. Using the tape measure, pencil mark 2'6" along one side, and 2' along other side.
4. Using a tack, string, and pencil, tack down one end of string, draw an arc from the 2'6" mark to 2' mark. The tack will not be in the corner but adjusted up 2'6" side.
5. Cut the 2x6 pieces in a curve using about three or four cuts with circular saw.
6. Cut the handles by notching the two center pieces.
7. Nail the crosspieces on the 2x6's with 8d nails to build the support frames.
8. Using one completed support frame, lay the frame on the outdoor carpet. With a felt tip pen, trace out the form of the frame on the carpet to be cut for the sides.
9. Flip to the other side and trace the other side on the carpet piece.

SHOPPING LIST

In two 5'9" pieces with two heights - 2' and 2'6"

All lumber should be pressure-treated.

- ✔Nine 2x6x8's for frame
- ✔Nineteen 1x4x12's for outer shell
- ✔10' x 12' outdoor carpet
- ✔6d and 8d decking nails
- ✔roofing nails

TOOLS NEEDED

- ✔Circular saw
- ✔Hammer
- ✔Carpenter's knife or shears to cut carpet
- ✔Two wood clamps
- ✔Tape measure
- ✔Carpenter's triangle
- ✔Framing square
- ✔Pencils
- ✔Felt tip pen

10. Stand the two end pieces up and nail the 1x4's on the bottom, front and back, with 6d nails.
11. Nail on the top 1x4 with 6d nails.
12. Slide in the center support frame, and nail down the 1x4's to it with 6d nails.
13. Nail on the remaining 1x4's, spacing carefully for an even fit.
14. Lay the carpet over the face and down the backside of the rolltop. With the felt tip pen, mark where the carpet is to be cut.
15. Lay the carpet on the floor (flat concrete surface or as level ground as possible), and use a long straight edge to mark the complete cut line on the carpet.
16. Cut the carpet with shears or a razor knife.
17. Lay the cut carpet on the rolltop, and nail down with roofing nails.
18. Nail on the side carpet pieces.
19. Repeat these steps from #14 for the second rolltop section.

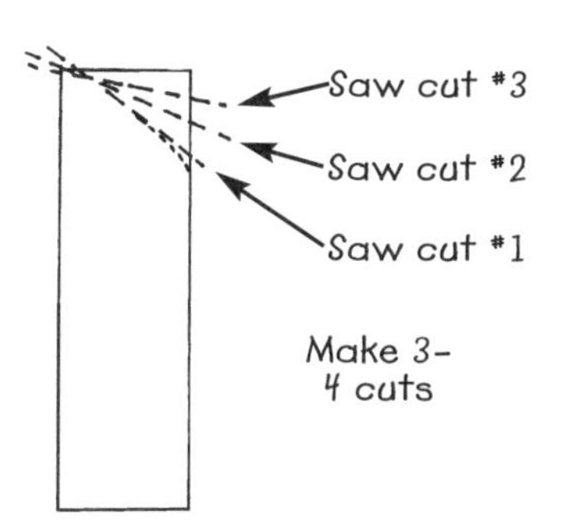

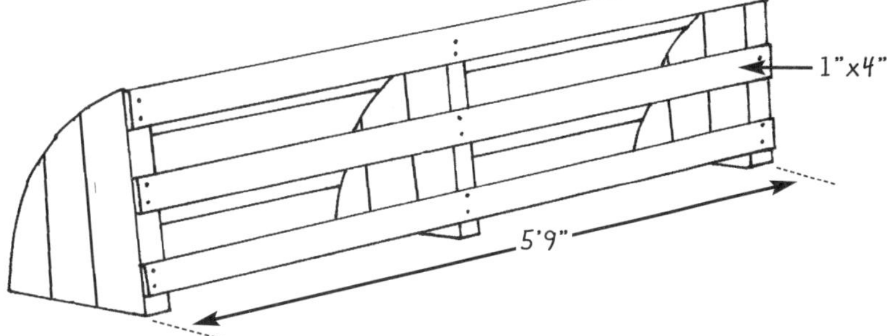

A front and side view of the roll top.

A back view of a roll top section before the carpet is attached.

Flower Boxes

A flower box filled with silk flowers

Construction Notes

1. Cut the four 2x6 inner support pieces as shown in the drawing (Four pieces for making two boxes).
2. Cut the 1x6 pieces to length.
3. Nail the 1x6's to the 2x6 support pieces using 6d nails.
4. Drill holes for the flowers using the 1½" wood boring bit, with spacing every 3 to 6 in. as desired.
5. Set the flowers in the holes. Plastic flowers are difficult to find, but silk flowers work well. To preserve their color, bring them in out of the weather when not in use.

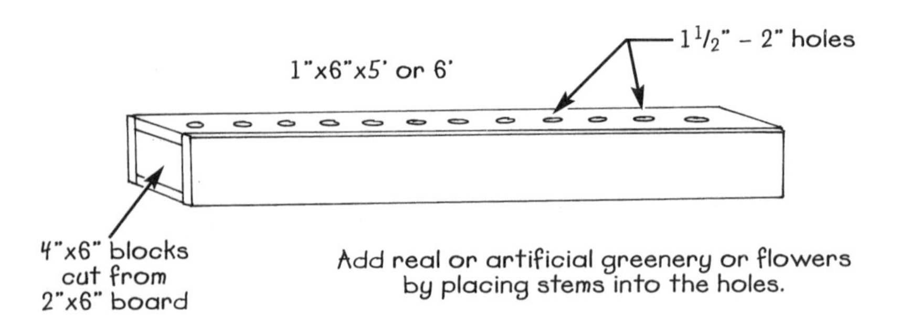

SHOPPING LIST

These plans will make a pair of flower boxes, 5' or 6' x 6". All lumber should be pressure-treated.

- ✔Four 1x6x12's for outer shell
- ✔One 2x6x8 for inner support
- ✔6d decking nails
- ✔Silk flowers may be purchased at a department or crafts store

TOOLS NEEDED

- ✔Circular saw
- ✔Hammer
- ✔Electrical drill
- ✔1½" wood boring bit
- ✔Tape measure
- ✔Carpenter's triangle
- ✔Pencils

Brush Box

The finished brush box ready to be stuffed with cedar or evergreen trimmings.

Construction Notes

Notes are written for a 10'x2' box. Adjust for a two piece box.

1. Cut all lumber to the specified lengths.
2. Angle cut the ends of the 2x6 base pieces.
3. Nail the base pieces to the two outer vertical 2x6 pieces with 8d nails.
4. Stand the two end pieces, and nail the bottom 1x6's on with 6d nails.
5. Nail the top 1x6's on with 6d nails.
6. Insert the center 2x6, and nail with 6d nails.
7. Nail on the center 1x6 pieces.

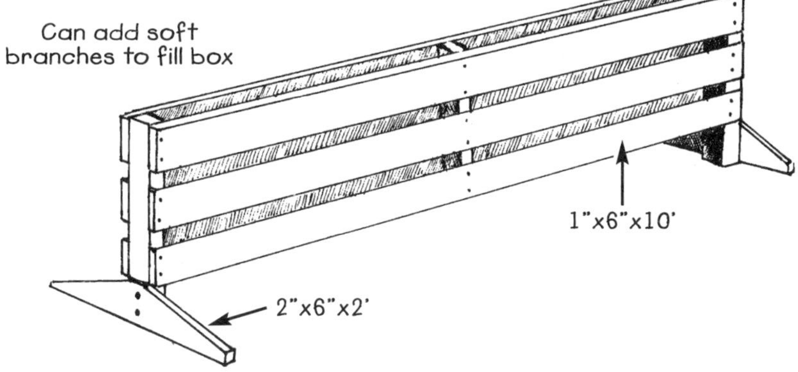

SHOPPING LIST

These plans will make either a 10'x2' brush box, or a brush box in two pieces of 5'9"x2'. All lumber should be pressure-treated.

- ✔Six 1x6x12's for the outer shell
- ✔One 2x6x10 for the base and inner support
- ✔6d and 8d decking nails

For two 5'9" x2' boxes

- ✔Twelve 1x6x6's
- ✔Two 2x6x8's
- ✔6d and 8d decking nails

TOOLS NEEDED

- ✔Circular saw
- ✔Hammer
- ✔Tape measure
- ✔Carpenter's triangle
- ✔Pencils

SHOPPING LIST

Untreated lumber is suitable for this project.

These planks can be painted as railroad crossing barriers as illustrated, or with any creative design. They can be used for logo jumps, too.

✔One 2x12x12

TOOLS NEEDED

✔Circular saw or jigsaw

✔Tape measure

✔Pencils

12' x 12" RR Crossing Planks

These planks are painted with railroad crossing stripes but can also be painted in other designs.

Construction Notes

1. Mark the jump cup handles as shown in the drawing.
2. Cut the ends out.
3. Optional: trim the inside joint with jigsaw.

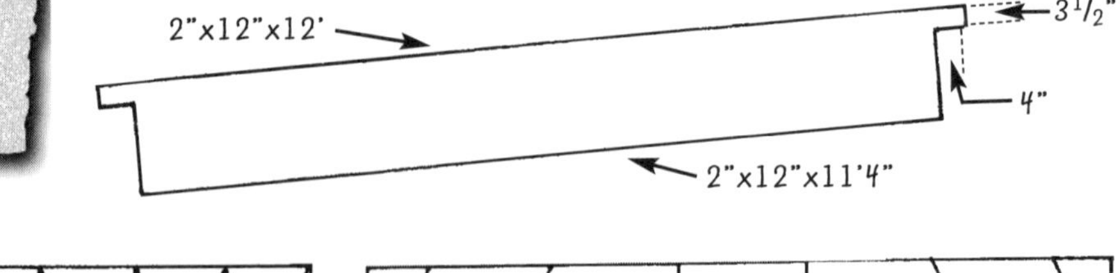

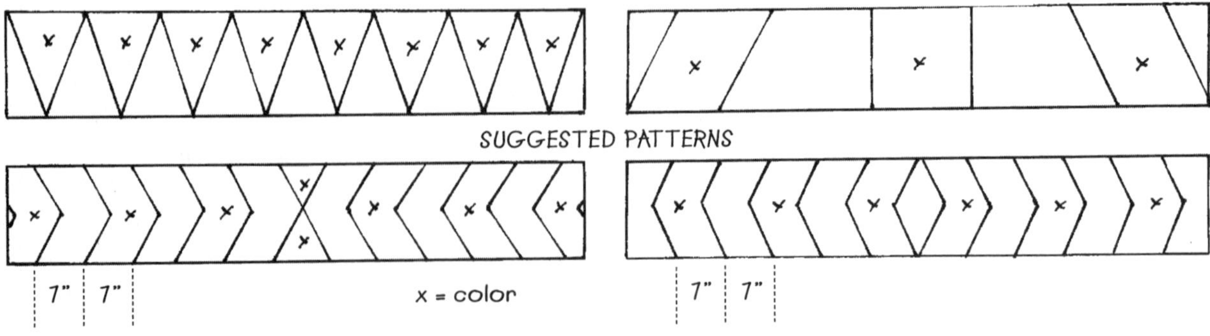

Mounting Block

Top: A side and back view of the mounting block.
Bottom: A front and side view of the mounting block.

SHOPPING LIST

Untreated lumber is suitable for this project.

- ✔ One 5/8 outdoor quality plywood sheet for the outer shell
- ✔ One 2x4x6 for the inner support
- ✔ 5d decking nails
- ✔ 4d bright finishing nails
- ✔ Two large sturdy door handles

TOOLS NEEDED

- ✔ Circular saw
- ✔ Hammer
- ✔ Jigsaw
- ✔ Electric drill
- ✔ Tape measure
- ✔ 12" ruler
- ✔ Framing square
- ✔ Phillips screw driver
- ✔ Pencils

Construction Notes

1. To conserve plywood, first cut a rectangle to 28 in. x 30 in.
2. Measure and mark cut lines as shown in the illustration to make the sides. These sides support the stairs.
3. Drill holes in the four corners large enough to insert the blade of the jigsaw.
4. Use the circular saw to cut the first two straight lines from either end. Then use the jigsaw to cut the other lines, using the drill holes to turn the blade in.
5. Cut the 2x4 into two 7 in. pieces and two 21 in. pieces.
6. Cut the lower and middle steps as two pieces of plywood of 10½ in. x 24 in.
7. Cut the two step facings of 6¼ in x 24 in. each.
8. Cut the top step of plywood 11 in. x 24 in.
9. Nail the 2x4 pieces to sides.
10. Nail the front and back pieces on, overlapping but flush with the sides. Nail into the 2x4 with deck nails, and into the plywood ends with finishing nails.
11. Nail on the steps and the top with finishing nails.
12. Nail on the facing with finishing nails.
13. Attach the door handles to the sides.

An underside view of the mounting block

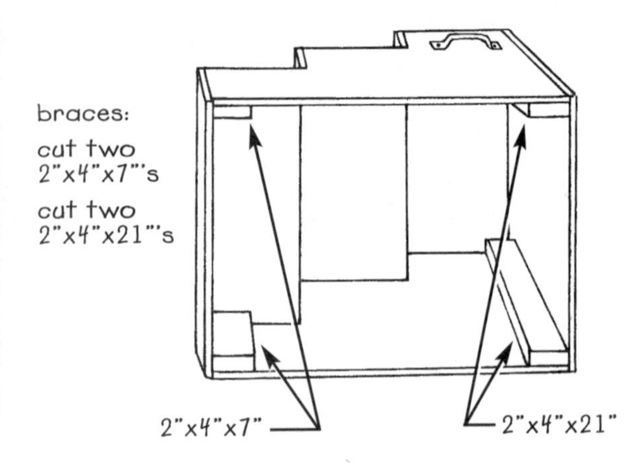

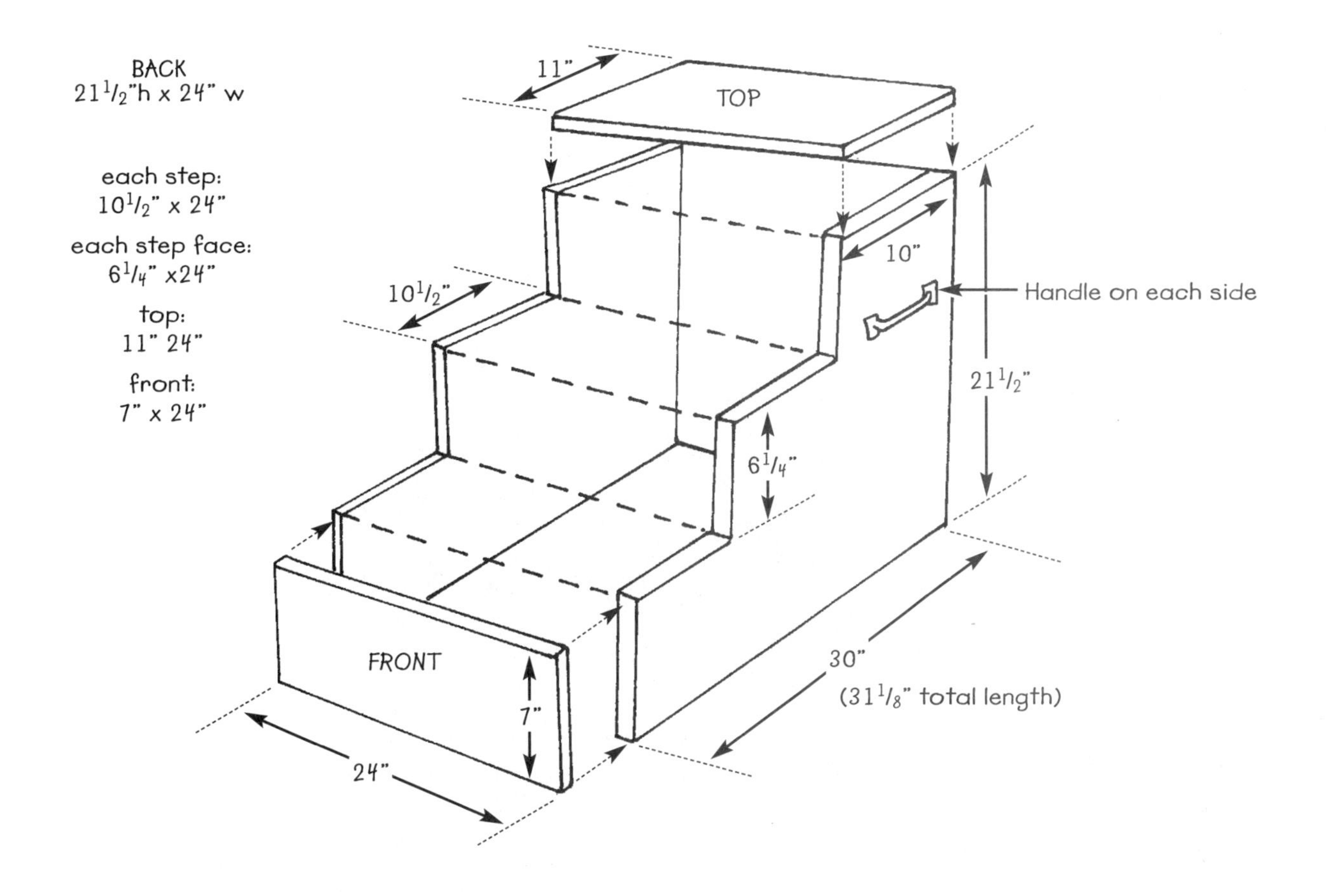

Measure out and mark cut lines. Drill holes in corners marked "drill". Cut first two straight lines to first drill holes. Then use jigsaw to cut other lines, using drill holes to turn blade in.

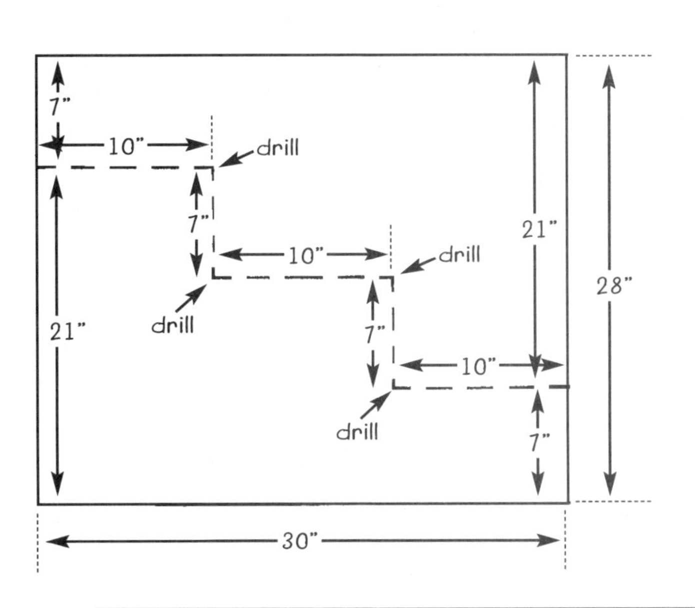

To conserve plywood, first cut rectangle to 28" x 30"

Sign Post

The sign post for the entrance to your farm or stable.

Construction Notes

1. Cut one 4x4 in half.
2. Bevel the four corners on both ends of one 4x4x4. (The other 4x4x4 will be discarded or used for another sign post.)
3. Bevel the four corners of one end of the 4x4x8. If you are planning to use a decorative top, this step is not necessary. The top may be nailed on when the post is finished.
4. Pencil mark the location on both 4x4's for notching the joints. Knowing the exact width is important when marking a 4x4 for notching. A full 4 inch wide notch will be too wide for making the sign post. Always use the lumber itself to mark the width for notching. Lay the 2x4 across the parallel 4x4's where it will be attached and pencil mark the width of the 2x4 for notching.
5. Notch the 4x4's. To notch a 4x4, first securely clamp the wood to a work table with the wood clamps. Set the depth of the saw blade to match the depth desired for the notch.

SHOPPING LIST

Use only pressure-treated wood.

- ✔Two 4x4x8's
- ✔Four 3x10 wood screws, phillips head
- ✔Optional: decorative top to post, pressure-treated

TOOLS NEEDED

- ✔Circular saw
- ✔Two wood clamps
- ✔Electric drill
- ✔#3 phillips drill bit
- ✔Tape measure
- ✔Carpenter's triangle
- ✔Pencils

From here, there are two methods to notch the wood the thickness of 4x4's. One is to make several parallel cuts between the pencil marks, then chisel out the remaining bits of wood. The drawback is that the chisel won't leave a smooth surface. The second method, and the one we use, is to first make cuts at the pencil marks. Then, make as many saw cuts as necessary to cut out all wood in between the two initial saw cuts.

6. Fit the notched pieces together.
7. Use the phillips head drill bit to screw in the wood screws to fasten the pieces together, two screws on each side.

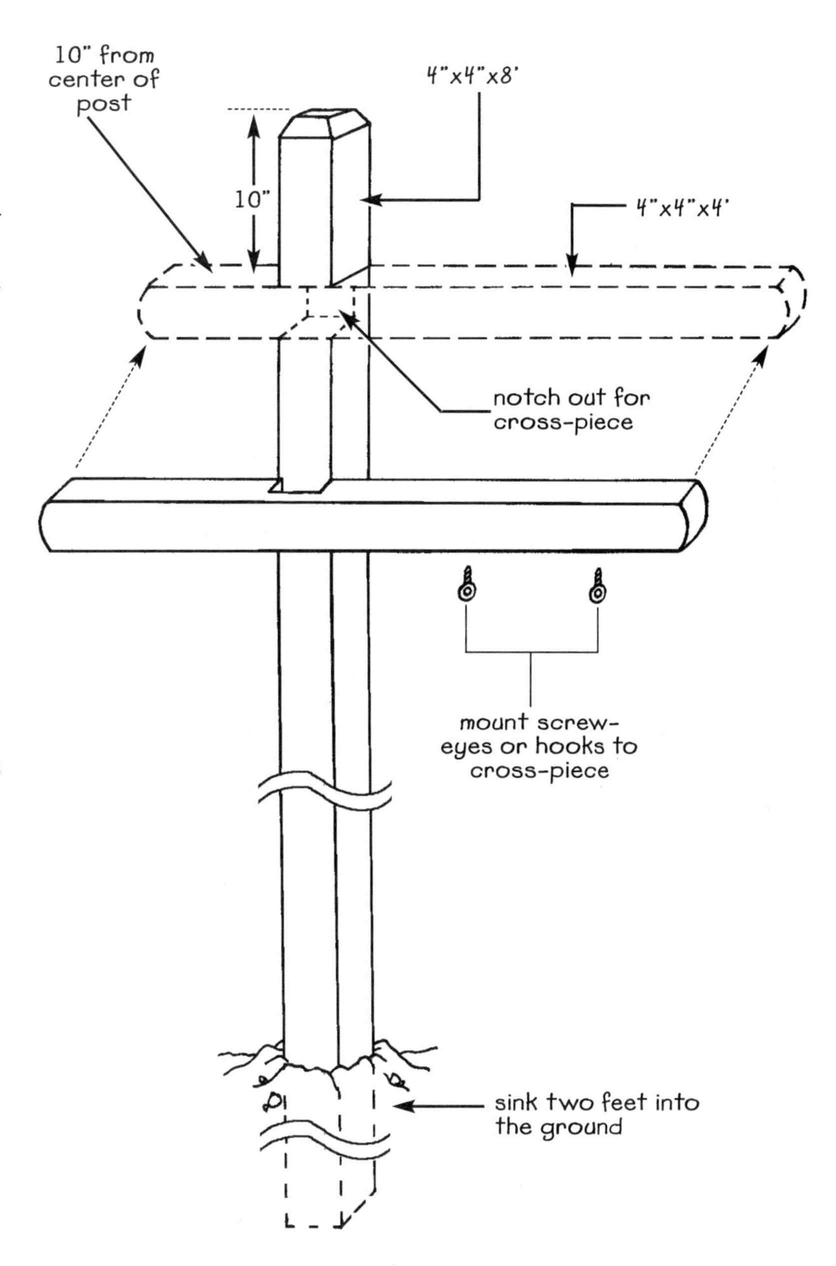

Saddle Stand/Storage Box

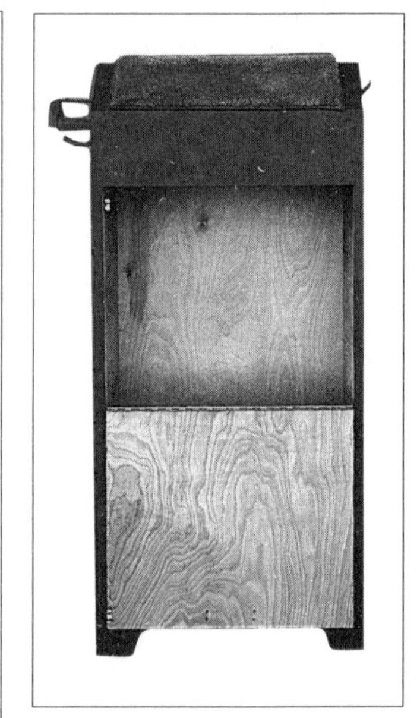

Left: The saddle stand/storage box is handy at your home stable, boarding stable or at a horse show.

Above: The saddle stand with the door open. Optional shelves or hanging hooks may be added.

SHOPPING LIST

- ✔One 1/4 or 1/2 in plywood sheet, smooth grain
- ✔One 18" piano hinge (these come in many lengths and are cut to size using a hacksaw)
- ✔Three door handles, two for carrying, one for the door
- ✔High quality wood glue (carpenter's glue)
- ✔One piece of carpet to go on top under saddle
- ✔Carpet tacks
- ✔Small door fastener

TOOLS NEEDED

- ✔Circular saw
- ✔Electric drill
- ✔Small drill bit
- ✔Two wood clamps
- ✔Tape measure
- ✔Screw driver
- ✔Jigsaw
- ✔Carpenter's triangle
- ✔Pencils

Construction Notes

1. Cut the plywood to the dimensions as indicated on plans. Make "legs" for the stand by cutting an arch on the ends of the plywood pieces that will be standing on the floor. A saddle stand with legs will be much more stable on uneven ground such as a dirt floor in a modest stable or the tack stall at a show. Mark the corners at 2½ in. as indicated on the plans. The top of the arch is 2½ in. off the floor. The inside corners may be curved or right angles. Pencil mark the lines and cut with the jigsaw.
2. Cut the door out of the front piece by drilling small holes in each corner just large enough to fit the jigsaw blade.
3. Attach the piano hinge to the front side and the door.

4. Assemble the pieces using wood glue or 3d bright finishing nails, the sides first, then the bottom, then the top pieces. The saddle stand shown is made using glue only for a smooth look.
5. Paint with base and top coats.
6. Attach the handles to the sides and the door.
7. Attach the door fastener.
8. Tack the carpet to the top with carpet tacks.

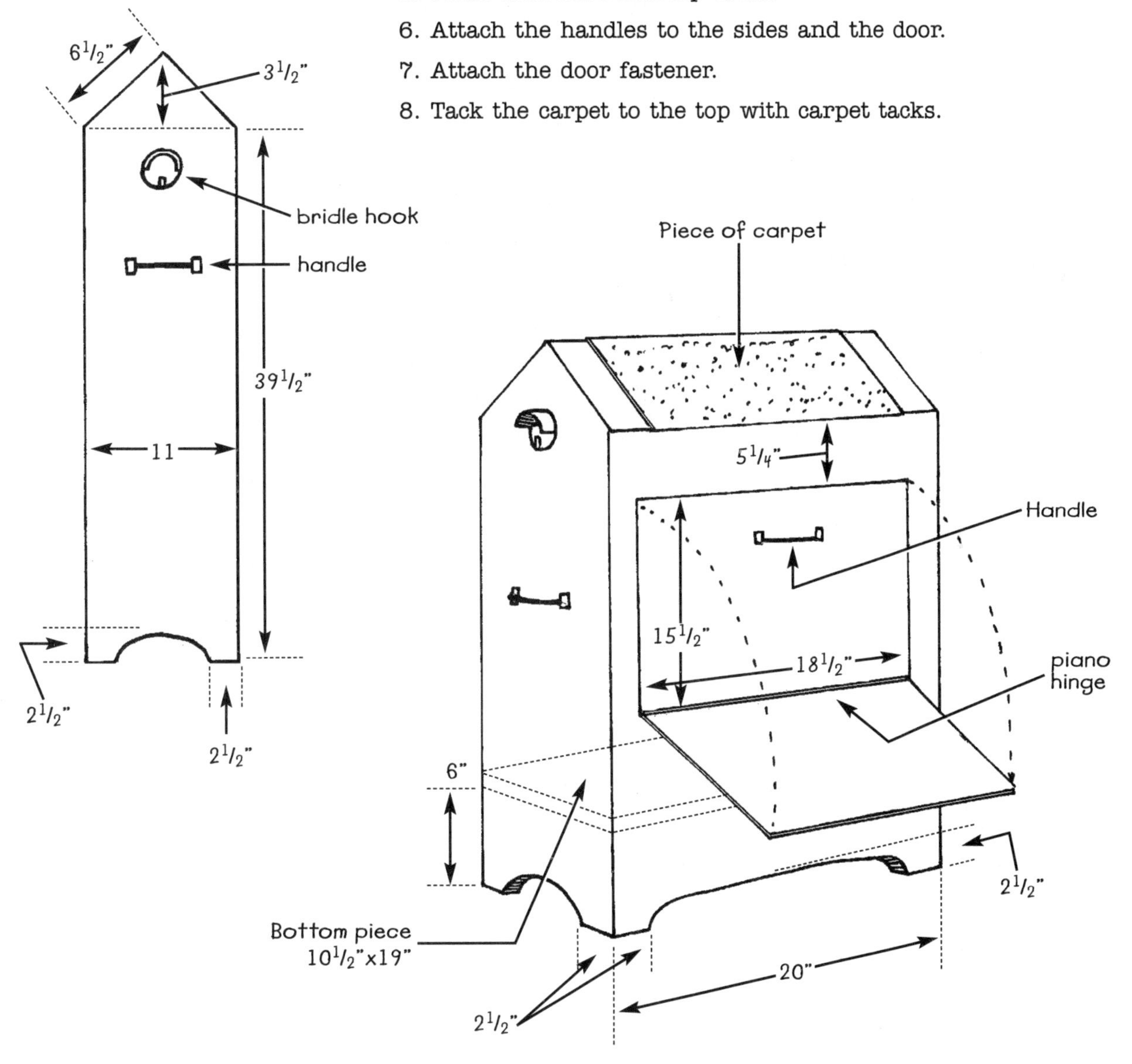

SHOPPING LIST

- ✔3" to 4" in diameter fence post or a broken jump pole
- ✔Two hook eyes
- ✔One double end snap
- ✔A piece of carpet remnant to cover the pole
- ✔Carpet tacks or roofing nails

TOOLS NEEDED

- ✔Circular saw
- ✔Hammer
- ✔Felt tip marker
- ✔Heavy shears or utility knife

Portable Saddle Rack

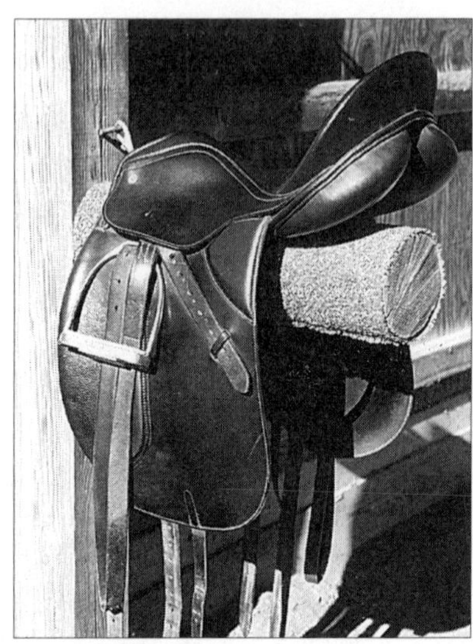

This handy portable saddle rack can be moved anywhere needed in the barn.

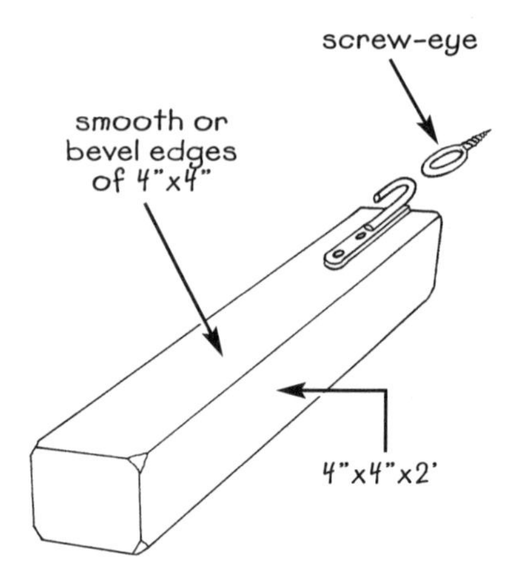

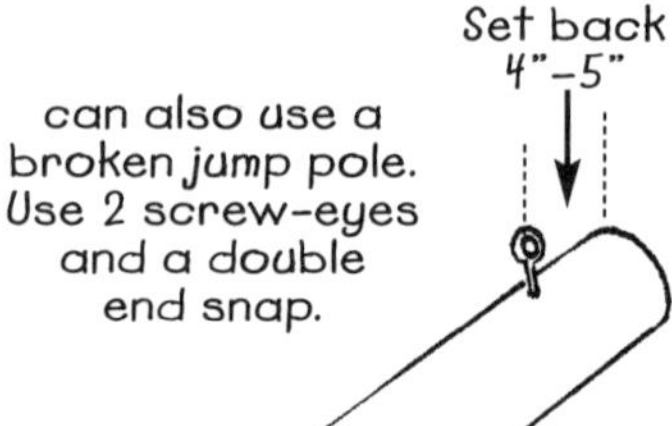

Construction Notes

1. Saw the pole to a 2½ ft. length.
2. Wrap the carpet around the cut pole piece inverted and mark where the carpet should be cut.
3. Cut the carpet to size.
4. Attach the carpet with tacks or nails.
5. Attach the hook eye to the pole.
6. Attach the other hook eye(s) to location(s) in barn where a saddle rack is needed. You can mount hook eyes by a stall, wash rack, grooming area, tack room, or as needed at horse shows in the tack stall.
7. Hang the saddle rack with a double end snap.

Grooming Box

Construction Notes

A grooming box can be varnished, painted in your stable colors or left natural.

1. Cut all the pieces of plywood for the four sides and bottom as indicated on the plans.
2. Angle cut the end pieces and drill ½ in. holes, 1 in. from the top.
3. Sand the edges of the wood pieces for a neat fit.
4. Assemble the four side pieces using the finishing nails or wood screws. If using wood screws, pre-drill the holes with a small drill bit, then use a screw driver to put the screws in.
5. Attach the bottom piece with finishing nails or wood screws.
6. Apply the wood glue to the drilled holes in the side pieces and insert the dowel.
7. Wipe off the excess glue.

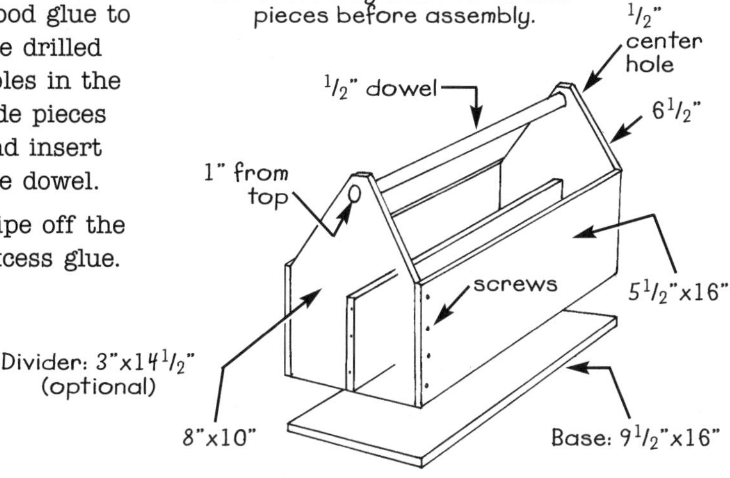

SHOPPING LIST

- ✔One sheet $^1/_4$" or $^1/_2$" plywood
- ✔One $^1/_2$"x16" wood dowel
- ✔3d bright finishing nails or wood screws
- ✔Wood glue
- ✔Medium grain sand paper

TOOLS NEEDED

- ✔Circular saw
- ✔Electric drill
- ✔Carpenter's triangle
- ✔Tape measure
- ✔$^1/_2$" wood auger drill bit
- ✔Pencils
- ✔Hammer (optional)
- ✔Phillips screw driver (optional)

Chapter 5

Paint Brightens The Ring

Bright, colorful jumps enliven the appearance a riding arena. Painted jumps also appeal to riders, inviting them to come jump the fences. If you run a boarding or training stable, you'll find that the color lends a polished, cheerful enhancement to your establishment and will often entice prospective clients to come visit your facility.

With pressure-treated, lumber you may opt for a rustic appearance and leave it unpainted. But painting cured treated lumber is an option, too. While bare, pressure-treated lumber is protected from bugs and subsequent rot, paint also serves as a preservative for both treated and untreated lumber. As long as untreated wood is painted and freshened every year or two, it will last just as long as the treated wood.

With pressure-treated lumber, the chemicals used in the treatment process are intended to prevent rot in wood that is in contact with the ground. This is because the ground almost always contains moisture, even when it seems dry. Untreated, unpainted wood will slowly absorb water. This will allow the wood to soften and be invaded by bacteria and mold which will further soften the wood. Then comes the armies of insects.

Pressure treatment of wood prevents the invasion of wood-softening agents, especially insects, but not of the water. Painting your jumps, whether treated wood or not, will effectively seal out moisture, allowing the wood to endure for years.

Painting treated wood is not without its problems though.

The chemicals may cause the paint to chip before the first year is out. Lumber salesmen will claim that treated wood will hold paint once the wood has dried, aged or turned gray. This may be true on some cases, but it is not always so. We have found over the years that aged, gray treated lumber may or may not hold the paint. There are many variables that go into creating a good bond between the paint and the wood which include temperature, humidity and the amount of chemicals in the treated wood. This goes for untreated lumber too.

> Basic tools needed to paint jumps vary just a bit to make the job go easier.

There is one steadfast rule, however: don't bother painting treated wood until it has dried. The moisture in the wood will prevent the paint from bonding and properly drying. Paint it too soon and the paint will wipe off in your hands.

How can you tell if the wood is ready? It does not need to be old and gray. In fact, once treated wood feels dry and shows cracks in the surface, it is probably ready for paint. Once your jump is built, cure it for painting by setting it outside. Curing may take just a few days out in the summer sunshine. In cold weather with little sunshine, curing may take one to two weeks. Don't bother setting it out if it is raining. Once surface cracks appear on all surfaces of the lumber, the jump is ready to paint.

Oil Versus Latex Paint

There are so many options to choose from when selecting the colors and type of paints, which also includes stains. Stain that is used on fences comes in white, black, red and brown. For a quick covering on just the poles and standards, stain may be a good choice.

For a brighter appearance, though, paint is a better choice than stain. Then there is the oil versus latex option to consider. Both have their advantages and disadvantages. The disadvantages to oil paint are: one, it needs about 24 hours to dry before another coat can be applied; and two, mineral spirits or other paint thinner is required to clean the brushes or sprayer.

For those pressed for time, latex paint may be preferable. The big advantage to working with latex paint is: **It's easy to clean up!** Just use soap and warm water, with no solvents needed. Depending on the air temperature and humidity, latex paint will dry within one to six hours, allowing you to apply the next coat fairly quickly.

Another difference between oil and latex paints is the texture of the paint once cured. Oil dries into a hard coating, which makes it more durable for trim work and floors in houses. Latex dries into a rubbery material; therefore the coating is softer. Since the only surface traffic that jumps encounter is the occasional rap of a shod hoof, the wood itself will chip whether it's painted with oil or latex in the final analysis. So, in our business, we always prefer latex.

When you paint your jumps, always plan on applying a base coat first, and then the top coat. Sometimes the wood will soak up more base coat and two top coats may need to be applied. The less expensive base paints, either oil or latex, tend to be thinner and soak into the wood more readily. Stick to quality paints. The finished product will look better, last longer, and will take less time to paint.

For the base coat, many paint dealers will recommend an oil base whether or not the top coat will be oil or latex. This is because, they say, the oil base will soak into the wood better, and the goal of the base coat is to set up a good bond for the top coat. But some dealers say a latex base soaks into the wood just as well as oil.

When painting, the ideal conditions are a warm air temperature with low humidity. Note the "warm," not hot. A word of caution here: do not paint jumps outside in the sun to expedite drying time. The sun will dry it too quickly, drying from the outside in and, therefore, not bonding properly to the wood. It will peel off in sheets in no time.

If possible, paint your jumps in the shade or on cloudy days. If the air is cool and moist or has a lot of humidity, it will take much longer to dry.

If you paint indoors, be sure to open the windows and doors. Use a box fan set on low to prevent the paint fumes from irritating your eyes and respiratory system. Without proper ventilation, you may find yourself with a headache, too. We always run a box fan when painting with either latex or oil. Disposable face masks are also available to block most of the paint fumes.

If you apply the paint with a sprayer, oil paint may be sprayed without thinning in most cases. Make a few tests before applying the paint.

Latex goes on much neater with a brush. To spray it on, latex needs to be thinned with water, otherwise it tends

Tools handy for painting jumps are the roller for large surfaces, a ½ in. brush for painting stripes or lettering, and a 2 in. brush for poles, pickets and gates. The edging tool is great for painting in the small spaces of pickets and gates.

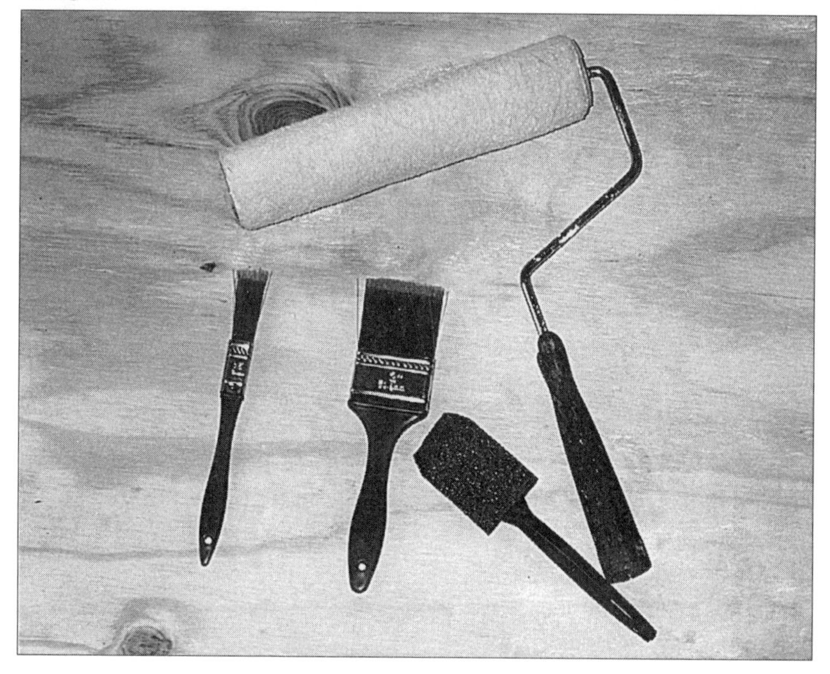

to go on too thickly. And when the paint goes on too thickly, it will run and not bond well to the wood.

Painting Tools

Basic tools needed to paint jumps and make the job go easier vary a bit. For painting poles, standards, gates, pickets, cavalletti, a 2 inch brush is the easiest to use. It covers plenty of surface and the brush can get into most of the corners. Certain brushes are labeled for the type paint you are using, either oil or latex. Getting the right brush for the right paint allows the paint to be spread evenly and efficiently

On large surfaces such as the coop, panel, or planks, a larger bush will save time. The trade off will be a heavier brush and your arm tiring more quickly.

For painting in between the pickets at the top of the jump and along the edges, an edge-painting tool works quite well. Trying to use a brush on these places will merely fray and ruin the brush for future use.

When painting stripes on poles or designs on panels, smaller brushes allow for neater lines. First prime the pole and put on a white top coat. To paint four colored and four white stripes on 10 ft. poles, make a stripe every 15 inches; for five stripes, make it every 12 inches. On 12 ft. poles, for four stripes make the stripe every 18 inches and for five stripes every 15 inches.

To make neat stripes, mark the pole for the desired number of stripes. Then use the carpenter's triangle to draw lines around the circumference. Using the color desired, paint neat lines along and fill in the stripe.

Masking tape does not work well to make neat lines unless the pole has been sanded to a fine finish. This is very time-consuming and probably not worth the time. When using tape on unfinished poles, the paint will bleed under the tape.

The same applies to painting the railroad crossing planks. Mark the lines with pencil, then neatly color in the lines and spaces.

Rollers can be used on large surfaces. The paint goes on a bit thicker, but a large surface can be covered quickly. On the brick wall, it will lend to the appearance of a textured surface.

Clean-Up

When using latex paints, drop the brushes or roller into a can or bucket of water until you are ready to end the painting project for the day. The water will loosen the paint and keep the paint from drying on the brush before clean up begins.

Wash latex paint brushes in running water to rinse out the paint. Warm water and soap help speed the process. Then place the brushes where they can get plenty of air circulation so they will be ready for your next project.

When using oil paints, drop the brushes in a can of paint thinner or mineral spirits. Rinse the paint from the bush and then squeegee out the remaining paint and thinner. Paint thinner is used to thin oil paints so if some residue is left on the brush for the next project, no harm will be done.

Specialized Painting

Schooling standards are usually painted white. A flat finish is quite appropriate and will be much more economical than gloss if painting several pairs of standards. Five to ten gallon buckets of flat white house paint cost quite a bit less per gallon than does white gloss exterior paint per gallon.

Semi-gloss or satin paints do well for colorful jumps. The contrast to flat paint will be lessened.

As a rule of thumb when building jumps to painted with colors or specialized artwork, buy untreated plywood, 2x4's and 1x4's that will not be in ground contact. The untreated wood does not contain the chemicals that treated wood does, therefore the paint will bond more securely. After putting in hours or paying someone to put on specialized paint, lettering or logos on a jump, you will not be in any hurry to see the paint chip off after a year due to chemicals in the wood.

Rainbows for Panels

Prepare the panel by applying the primer and top coats. To make a rainbow pattern, use a string with a pencil tied on one end and a nail on the other. Find the center point at the bottom of the panel. Depending on how deep or shallow the arcs are to be, the center of the arc will either be at the bottom of the panel or several inches below that center point off the panel. Gently tap the nail into the center point, let the string out and make your pencil marks on the painted panel. Repeat as necessary to draw parallel arcs to complete the rainbow.

The number and thickness of the arcs and the color scheme can be of any dimension you choose. The same method can be used to make a target or a sunburst. This is the time to be creative!

To Paint a Brick Wall

Prepare the panel (the coop, plank, etc.) by filling in any cracks with plastic wood filler. Fill in the seam between the two pieces of plywood on the panel and

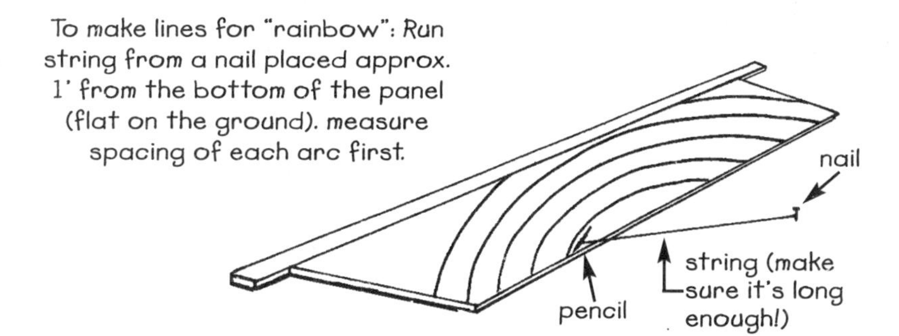

This panel has arcs made in a narrow pattern. Many variations are possible.

along the top and bottom edges of the plywood. This will seal out rainwater. If left open at the top, rainwater can enter the plywood, causing the surface eventually to swell and buckle. Once the wood filler is dry, gently sand any rough edges smooth.

Next cover the whole panel both sides, top and bottom with primer paint. The panel can be hung in jump cups on standards to do the entire panel in one session. If not available, lay the panel on saw horses and do one side at a time. When dry, completely recoat the panel with a white top coat. Once that is dry, it is ready for brickwork.

To make the brick pattern, you will need a roll of ¾ in. masking tape. Measuring and taping the brick pattern will take time, but the effort will worth it and you will be proud of the handiwork.

The tape will mask the white paint which becomes the "mortar" in between the red or brown bricks. The average brick on your house measures roughly 2x8 inches. In this example, we will use similar dimensions but expanded to evenly fit the panel.

Begin by making pencil marks down the plywood section every 4 inches on both side edges of the plywood. Then make at least 3 or 4 marks across the plywood. Run a strip of masking tape across the plywood immediately under the 1x4 or 2x4 top piece, then run another strip across immediately above the bottom 1x4. Then at every mark, run strips of masking tape horizontally across the plywood, just above the pencil marks. This marks your rows of bricks.

For the next step, mark across the horizontal rows every 9 inches. Each 9 inch space will encompass one "brick" and one strip of "mortar." Alternate on each row, with the starting point from the side to be 9 inches on one row, 4 1/2 inches the next, 9 inches on the next and so on.

Next mark the top and bottom 1x4's (or 2x4) every 9 inches, repeating the alternating pattern on the plywood. Put a strip of masking tape vertically just to the left of every pencil mark on the plywood, 2x4 and 1x4.

When the taping is complete, paint the entire side with a red brick color paint. It is best to paint the brick while the panel is laying flat and not hanging. This will prevent the paint from running onto the "mortar" as the tape is pulled off. Some paint may bleed under the tape and will give the brick a rough natural appearance.

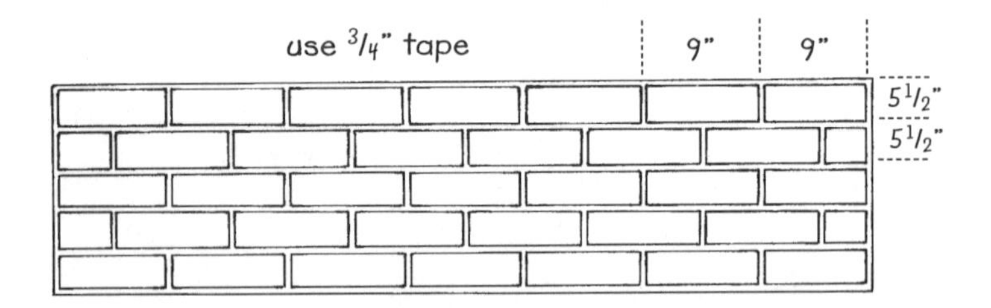

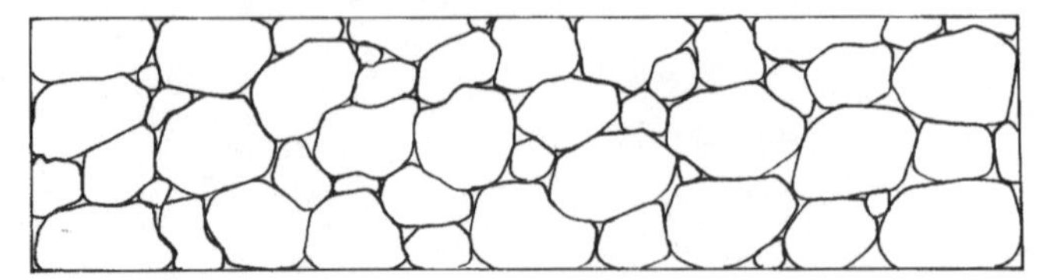

If using latex paint, begin removing the tape as soon as you have completely covered the panel, before the paint has dried. If the tape is left on until the paint has completely or partially dried, the dry rubbery latex will tear as the tape is pulled off. This will break the paint's bond to the surface, leaving the edges of the "brick" vulnerable to the weather when left outside in the future.

Once the masking tape has been completely stripped off, you are finished with side one. And there stands a realistic looking brick wall.

You can leave the other side a blank white panel to be slanted in the jump cups to simulate a coop. Or you can paint anything, such a business, farm or show logo and lettering, a stone wall, a sunburst, pumpkins, anything that comes to mind.

Chapter 6

Logo Jumps For Advertising, Business, Farm or Show Names

Logo jumps are increasingly popular for two reasons. One is that it is an attractive way for show or class sponsors to display their product or company name. The specialized jump can be moved from show to show, or it can stay on the same show grounds.

The other reason is that it can be used to provide useful photo opportunities. One thing that brings competitors back to a show are good memories. Professional photographs received after a horse show enhance fond memories of a job well done. The photo then becomes a record of that specific show, not just another pretty jump picture.

From a business standpoint, for the show management or show sponsor, a great jump for photo opportunities are panels that have the organization, farm, or show name and logo on the off side. On the near side, the horse and rider see a brick panel on their approach. On the off side, the photographer catches them coming over the jump, recording the name of the show in the photo.

Other kinds of jumps can be successfully used to display a logo too. Any jump with a broad flat surface such as a coop or planks will work. One feed company even put their name on a brush box painted in the company colors on one side. The other side remained plain white and was used at local hunter and combined training events. The two piece brush box was easy to transport in a small pickup to the many local shows.

Another option to display a logo is to mount it on wing standards. Once the wing frame is built, substitute a

22 in. by 28 in. plywood panel for the 1 in. x 4 in. pickets and paint the logo on it.

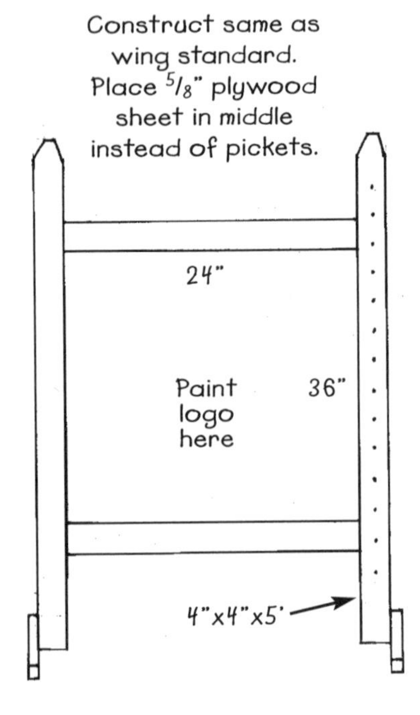

Putting the Logo on the Jump

To help the logo paint job to look its best and last the longest, the bare wood surface should be smooth before applying the primer coat. Sand it if the wood surface is rough and use a plastic wood filler to fill in cracks to create a smooth surface. The lettering and logo will look much neater and be easier to paint on. This is especially important if a professional sign painter is hired to do the job.

Plan on using a high quality polyurethane paint for the logo and lettering. This way your hard work will endure the elements for some time to come.

Pony Club member Jennifer Wilson jumps their logo panel.

If painting indoors, be sure to ventilate the room adequately. First, paint both sides of the assembled panel (or other jump) with the primer. When completely dry, cover both sides with the top coat. The side to be painted with the logo can be painted with a color to match the farm or company color scheme. When dry, the panel is ready for the logo.

The simpler the business, show or farm logo is, the easier it is for us unartistic-types to copy it onto the panel. The easiest way to do it is to enlarge the logo to an appropriate size for the panel, and tack or tape it on the painted panel to trace around it. If the logo sample is not big enough, take a clean copy to the nearest copier machine that enlarges, at the office, library, copy shop, or post office.

Once you have a copy of the logo of a suitable size for the panel (or other jump), the next step is to plan the placement of the logo and letters on the jump. If it is simple enough and large enough, cut out the logo and lay it on the panel. Use small pieces of tape to anchor it to the panel in strategic places to prevent shifting. Next trace around the cutout in pencil. When finished, pull off the cutout and clean up the pencil lines.

Lettering stencils for the show or farm name can be purchased at an office supply store and some hardware stores. Try out the stencils by tracing the lettering on a large piece of cardboard first to become familiar with proper spacing of the letters. Next, draw a pencil line on the panel that the letters will "rest" on. This is to ensure the letters are at the same height. Tape the stencils on as with the logo and trace the letters. Remove the tape and clean up any pencil lines as necessary. Your last step is to neatly paint the logo and letters.

For those less confident of attempting the lettering, or if the logo is more complicated artistically than a mere silhouette, your organization may consider hiring a sign painter to do this side of the panel for a neat and professional job. Remember, the panel will be an advertisement. It should neat and well done.

When hiring a professional to do the logo, you can save money by painting the panel yourself after building it. Apply the primer, top coat for the brick (or stone or white side) and the top coat on the logo side in the color you want. Make arrangements with the sign painter to paint on the logo and lettering in the color needed and deliver the panel to the sign shop.

At the Show

At the show, hang the panel in flat cups for the safety of the horses and riders should a jumping mistake occur, and to prevent damage to the panel. The pole used over top the panel may be hung in regular cups. If the course has been designed with the panel and photographs in mind, be sure to discuss it with the show photographer before the event begins.

Show managers and course designers can play a big part in making their show memory jump attractive to both the photographer, the horse, and the rider. It's important to place the jump in a suitable location on the jump course and to decorate it tastefully to the eye with potted plants or flowers, but don't overwhelm the horse in the process! The photo fence should be as well decorated on the off side as any other jump in the ring.

With these things taken into consideration, the end result will be a beautiful photograph of the horse jumping in great form, over a colorful fence with the show's name, logo, or sponsor emblazoned on it. And the rider will more inclined to buy the photo, because the show is identified by the attractive jump, and the horse looks like a star. It won't be just another show photo.

Wilson jumps from the brick side of the panel.

Chapter 7

Setting Up A Dressage Arena

Carol Noggle, U.S. Pony Club Vice Regional Supervisor, once said, "Probably few people riding dressage tests know what goes into setting up a dressage arena."

How right she was! The basic difficulty people have is knowing how to make the corners of the rectangle with perfect right angles. Because if they aren't perfect right angles, the ring will end up as a parallelogram, and a bit difficult to ride.

The key to setting up a proper arena is knowing the correct distance for the diagonal. Dressage makes practical use of the mathematical Theorem of Pythagoras:

a + bb = cc

For mathematic and history buffs, Pythagoras proved the theory in the 6th century b.c. and it was used by the Egyptian surveyors.

A large arena is 60 meters by 20 meters. Use the above formula to figure the diagonal:

(20x20) + (60x60) = cc
400 + 3600 = 4000
square root of 4000 = 63.2
the diagonal is 63.2 meters

A small arena is 40 meters by 20 meters. The formula is:

(20x20) + (40x40) = cc
400 + 1600 = 2000
square root of 2000 = 44.7
the diagonal is 44.7 meters

Determine the Arena Location

The first step is to walk the area designated for the dressage ring with a critical eye. Where are the dips, the potholes, the hard, slick spots, the mud puddles, the boulder in the ground, and/or the downhill slope? These obstacles should be avoided in the horse's path on turns, circles and along the rail if possible.

You don't want unbalanced green horses struggling through the turns.

If it is to be outdoors in a field, find the most level spot available. Where is the judge going to sit all day in relation to the sun? If it is to be indoors, look up at the ceiling. Many indoors have the spine of the roof over the approximate centerline of the arena. This is a visual reference judges and riders may use unconsciously to determine the placement of the centerline. The long sides should be parallel to the walls, another visual reference point.

Once the arena is measured off, the materials that can be used to border the ring may vary from simple plastic chains to attractive plastic rails purchased from dressage catalogs. A friend who regularly puts on dressage schooling shows constructed an arena border using cinder blocks standing on end with painted 2x4's inserted in the block openings.

Plastic chains often used to decorate yard and sidewalk borders are fairly simple to erect and inexpensive. The chains are hung on the metal posts with shallow loops. The posts are metal stakes with decorative caps that not only look nice, but keep the chain on the post.

Choose the right type of hammer for driving in the stakes. The best type hammer is a short-handled hammer with a large head. A regular carpenter's hammer (or claw hammer) is too lightweight, and will require strength and endurance to drive in all the stakes needed for a full ring. A sledge hammer drives in the stakes well but, again, requires strength and endurance for a full arena's worth due to its weight.

Torrance Watkins rides Lou Llewellyn in the dressage phase of the advanced division at Morven Park Horse Trials. The arena border consists of plastic chains set on metal stakes. The plastic caps hold the chains in place.

There is some variation in the kind of stakes to be used too. Some work well in almost any type of ground. Long stakes tend to be stable in soft ground, but can be difficult to drive into hard ground. Short stakes do nicely in hard ground and need less pounding, but tend to fall over in soft ground.

In very hard ground, driving in a railroad-type spike a short ways, then pulling it out, and driving in the ring stake can help. Be sure to not drive in the spike too far or it will be difficult to pull back out.

Another advantage of using a plastic chain type arena is that can be erected alone.

Some of the more elaborate arena borders require at least two people. When using the chain border, leave a bit of string or baling twine with the scribe or ring steward for emergency repairs if a horse should break it.

The advantage of the cinder block/2x4 rail arena is that it requires no pounding of stakes. The same applies to the plastic pylon and rail sets commercially available.

With the arena border materials on hand, it's time to measure out the arena. The tools required are two metric tape measures that do not stretch and four relatively short metal stakes. One measure tape should be 100 meters and the other can be 60 meters. These tapes are available from certain dressage arena suppliers. They can also be purchased from land survey companies for about $1 per meter (or foot).

While standard 100 ft. measure tapes can be used and are cheaper, all lengths must be converted from feet to meters (3.28 feet = 1 meter). This leaves room for cumulative errors if you're not careful. By far the simplest and most accurate is using metric tapes.

When measuring off an arena, three people working together is ideal, but it can be done with two as well. When setting up more than one ring, four people are ideal. Everyone can help measure out the first arena. Then two move on to measure out the next. The other two can then set up the arena border and letter in the first.

Use the accompanying drawing for measuring using a 60 meter tape and a 100 meter tape:

1. Decide the general placement of the arena.
2. Drive a stake securely in the ground at Corner (1).
3. Using a 60 meter tape, measure 60 m (40 m, small arena) down the first long side.
4. Pull the tape in a straight line, shake out the curves and slack, pull the tape tight and lay it on the ground. Mark 60 m (40 m, small) to make Corner (2), drive in a stake.
5. This line will not shift and is the base for all remaining measurements. Be sure the line is where you want it, with the footing and placement in relation to the judge's stand, wall, fences, trees, rocks, mud puddles, etc. Now is the time to adjust the line.

A dressage arena with a border of cinder blocks and painted 4x4 rails. The rider is the author on Black Poppy.

6. Go back to Corner (1). Using the 100 meter tape, secure it to the stake, measure 20 m across the short side to Corner (3). Drive in a temporary stake at Corner (3), and run the tape around the stake or have a helper hold the tape at the 20 m mark.
7. Using the 100 meter tape still attached at Corner (1), measure to Corner (2) across the diagonal 63.2 m (44.7 m, small), or to 83.2 m (64.7 m, small) on the tape because 20 m + 63.2 m = 83.2 m. Secure the tape to the stake at Corner (2) or have a helper hold the tape.
8. Move the Corner (3) stake as needed to draw the tape tight and straight to realign it into a right angle and form a tight triangle.

The corners of the dressage arena diagrams are numbered in order of the steps explained.

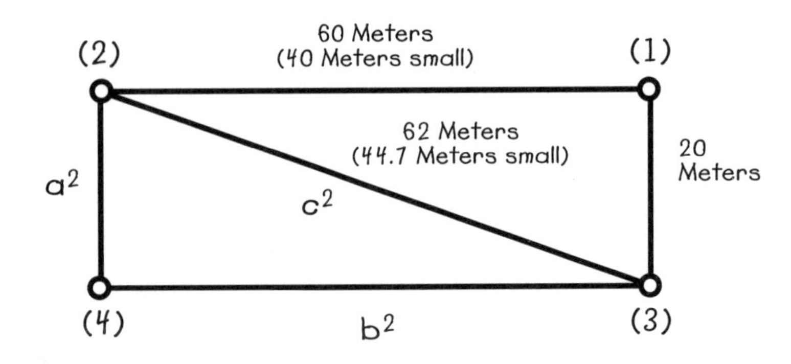

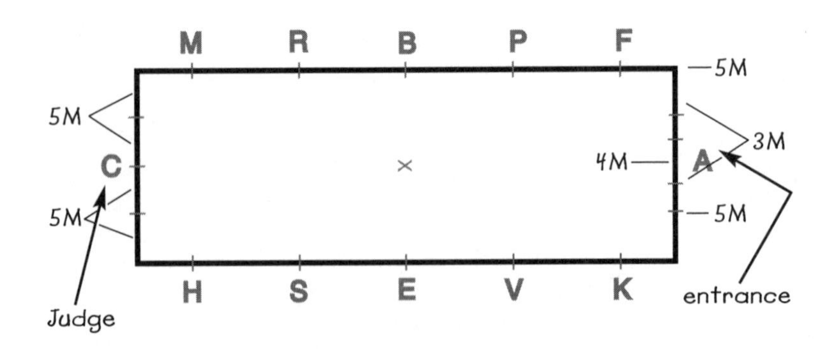

Judge(s) at 5 meters and raised 0.5 meters

Public at 20 meters

9. Move the temporary stake at Corner (3) and drive it in at realigned Corner (3). The distance from (1) to (3) should be exactly 20 meters.
10. Remove the tape running across the diagonal from the stake at (2). Find the 100 m mark on the tape and secure or hold it at (2).
11. Find the 80 m mark on tape. Pull the tape tight to form a right angle at Corner (4). Drive a stake in Corner (4), and lay the tape on the ground. Your perfect rectangle with four square corners is now laid out.
12. Lay stakes out on the long side at every 6 meter mark on the tape. On the short side with A, lay stakes at 5 meters, 3 meters, 4 meters (at entrance), and 3 meters. On the short side with C, lay stakes every 5 meters.
13. Drive in stakes at all locations.
14. Lay out and attach chain or other border.
15. Set the letters a half meter out from border in proper order.

Use an arena diagram or remember "All Fat Black

Mares Can't Hardly Even Kick." Place the letters counter-clockwise, beginning at the entrance. **M**, **F**, **H**, and **K** are 6 meters from the corners, and the other long side letters are spaced 12 meters apart. **A** marks the entrance and proceed from there. Then, starting between **B** and **M**, and going counter-clockwise, remember RSVP.

Letters for Dressage Arenas & Other Uses

Dressage letters and other show information signs can be made from many materials. Two simple types are made from plywood. These can be of any thickness, even the ¼ inch variety, but preferably of an outdoor quality.

In the photos at the right, the plywood is cut to 18 in. across by 21 in. in height. Cut the plywood pieces to the size needed. For the stand-up leg, cut a 1x2 to 21 in. Attach to the top back of the board with a cabinet hinge.

The letters may be painted after the initial base and top coats of paint are applied. In the photo, wood letters were purchased at a home supply store and glued on with wood glue. The letters are then painted a contrasting color. Letters can also be painted on using a stencil.

In the bottom photos, plywood was used to make cross-country markers indicating the half-way point. The important aspect of this board's design is that it won't blow down in the wind. The same design can be used for dressage letters, ring numbers or any other need.

The board in the photo measures 14 in. across by 24 in. in height. The base and top coats of paint were applied, then the lettering added. The key element here is a piece of iron rebar cut to 24 in. long. Mount two metal brackets on the rear of the board leaving enough space for the rebar to slip through. At the show, the rebar can be pounded into the ground until flush with the top of the board.

Top Left: An arena marker made of plywood with a wood letter glued on.
Top Right: A side view of the arena marker
Bottom Left: Another type marker used to mark the half-way point on a cross country curse.
Bottom Right: The back view of the half-way marker with brackets for a piece of iron rebar to slide through and sink into the ground.

Recommended Reading

Books on Course Design and Training Over Fences To Help Decide What Jump Equipment to Build

AHSA Rule Book, American Horse Shows Association, New York.

Cavalletti, Reiner Klimke, J.A. Allen, London, 1986.

Designing Courses & Obstacles, ed. by John H. Fritz, Houghton Mifflin Co., Boston, 1978.

Practical Eventing, Sally O'Connor, Half Halt Press, Boonsboro, Md., 1998.

Schooling Your Horse, Vladimir Littauer, Arco Publishing, N.Y., 1982.
This book is out of print, but you may obtain it from a used book dealer, or borrow it from your local library or through their inter-library loan service.

Safety Checklist

- [ ] When lifting *anything* (including jumps), always bend at the knees and keep your back straight.

- [ ] Use wood clamps to secure the lumber when drilling or cutting. Clamping the wood and using two hands for running the drill or saw will make for safe working conditions.

- [ ] Read the information manual that comes with your power tools cover to cover. And, if you haven't used your tools in a while, review the manuals before using.

- [ ] For personal protection, always wear safety glasses while sawing, drilling and hammering.

- [ ] Wear a dust mask when sawing or drilling, especially on treated wood. Whenever possible, do your drilling or sawing outdoors, or at the very least, in a well-ventilated area with a fan running.

- [ ] After handling uncured treated lumber, always wash your hands to get rid of the chemicals (similar to arsenic!) that has been injected into the wood. Use gloves when loading or unloading lumber, but NEVER when sawing or drilling. They can too easily get caught in the tool.

- [ ] Wear sturdy shoes: a piece of lumber dropped on your foot is not pleasant!

- [ ] When using power tools never wear loose clothing. Tuck in your shirt tails, roll up loose sleeves, and tie back long hair.

- [ ] Try to do your painting outdoors if possible. Dust masks for painting are advised.

- [ ] Do not work with power tools when you are tired. Studies show that's when most accidents (of all kinds) occur.

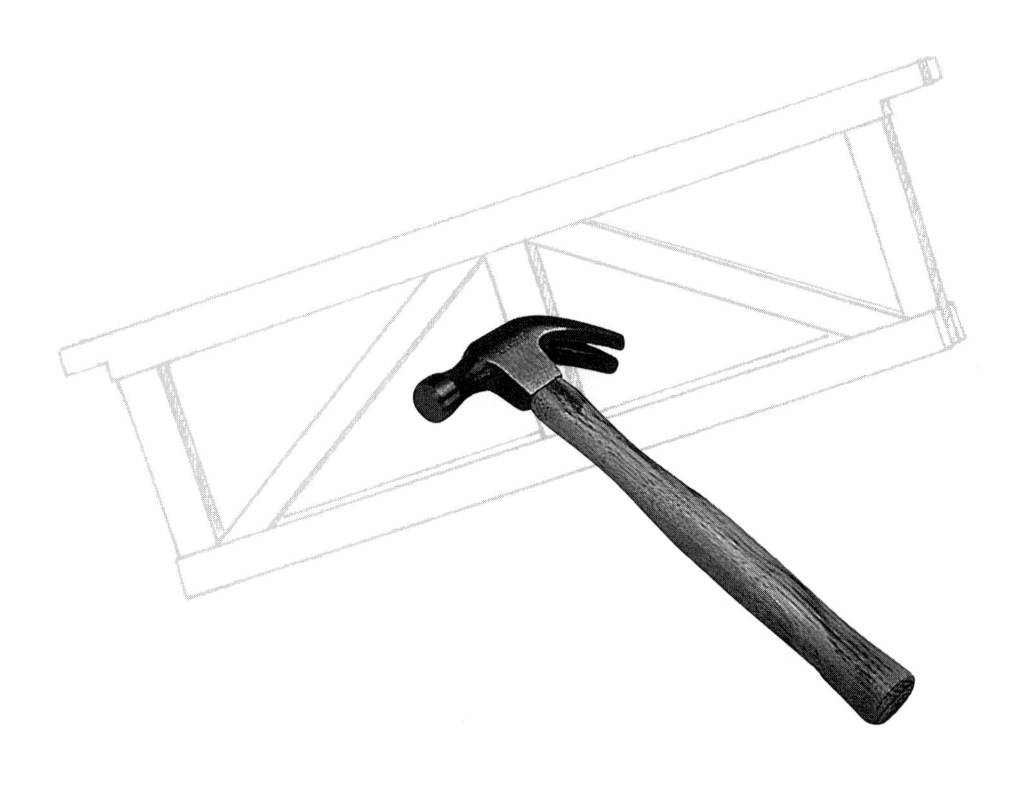